Computational Error and Complexity in Science and Engineering

This is volume 201 in
MATHEMATICS IN SCIENCE AND ENGINEERING
Edited by C.K. Chui, *Stanford University*

A list of recent titles in this series appears at the end of this volume.

Computational Error and Complexity in Science and Engineering

V. Lakshmikantham
FLORIDA INSTITUTE OF TECHNOLOGY
DEPARTMENT OF MATHEMATICAL SCIENCES
MELBOURNE, FLORIDA

S.K. Sen
FLORIDA INSTITUTE OF TECHNOLOGY
DEPARTMENT OF MATHEMATICAL SCIENCES
MELBOURNE, FLORIDA

2005
ELSEVIER
Amsterdam – Boston – Heidelberg – London – New York – Oxford
Paris – San Diego – San Francisco – Singapore – Sydney – Tokyo

ELSEVIER B.V.
Radarweg 29
P.O. Box 211, 1000 AE Amsterdam
The Netherlands

ELSEVIER Inc.
525 B Street, Suite 1900
San Diego, CA 92101-4495
USA

ELSEVIER Ltd
The Boulevard, Langford Lane
Kidlington, Oxford OX5 1GB
UK

ELSEVIER Ltd
84 Theobalds Road
London WC1X 8
UK

First edition 2005

Library of Congress Cataloging in Publication Data
A catalog record is available from the Library of Congress.

British Library Cataloguing in Publication Data
A catalogue record is available from the British Library.

ISBN: 0-444-51860-6
ISSN (Series): 0076-5392

∞ The paper used in this publication meets the requirements of ANSI/NISO Z39.48-1992 (Permanence of Paper).
Printed in Great Britain.

Preface

The monograph focuses on an estimation of the quality of the results/outputs produced by an algorithm in scientific and engineering computation. In addition the cost to produce such results by the algorithm is also estimated. The former estimation refers to error computation while the later estimation refers to complexity computation. It is mainly intended for the graduate in engineering, computer science, and mathematics. It can also be used for the undergraduate by selecting topics pertinent to a given curriculum. To gain practical experience, any such course should be supplemented with laboratory work.. Besides, it would be of value as a reference to anyone engaged in numerical computation with a high-speed digital computer.

If we have to compare two or more algorithms to solve a particular type of problems, we need both error and complexity estimation for each of the algorithms. Whenever we solve a problem and produce a result, we would always like to know error in the result and the amount of computation and that of storage, i.e., computational complexity and space complexity. The monograph is precisely an exposition of both error and complexity over different types of algorithms including exponential/combinatorial ones.

Chapter 1 is introductory. It discusses the distinction between science and engineering, highlights the limitation of computation, tools and types of computation, algorithms and complexity, models of computation, computer-representable numbers, and stages of problem-solving.

Chapter 2 is an exposition of all that is connected with error. Precisely what error is, why we get error, and how we estimate the error constitute the core of this chapter. Similarly, Chapter 3 explains what, why, and how of complexity of algorithms including various types of complexity.

Errors and approximations in digital computers constitute Chapter 4. The details of IEEE 754 arithmetic are also included in this chapter. Chapter 5, on the other hand, presents several numerical algorithms and the associated error and complexity.

Error in error-free computation as well as that in parallel and probabilistic computations are described in Chapter 6. The confidence level which is never 100% in probabilistic computations is stressed in this chapter.

Simple examples have been included throughout the monograph to illustrate the underlying ideas of the concerned topics. Sufficient references have been included in each chapter.

Certainly a monograph of this type cannot be written without deriving many valuable ideas from several sources. We express our indebtedness to all the authors, too numerous to acknowledge individually, from whose specialized knowledge we have benefited.

V. Lakshmikantham
S.K. Sen

Contents

Chapter 1

Introduction

1.1 Science versus engineering

The Collins Gem dictionary meaning of science is the systematic study of natural or physical phenomena while that of engineering is the profession of applying scientific principles to the design and construction of engines, cars, buildings, or machines. All the laws of physics such as the Newton's laws of motion, the first and second laws of thermodynamics, Stokes law, all the theorems in mathematics such as the binomial theorem, Pythagoras theorem, fundamental theorem of linear algebra, fundamental theorem of linear programming, all the laws, rules, and properties in chemistry as well as in biology come under science. In engineering, on the other hand, we make use or apply these rules, laws, properties of science to achieve/solve specified physical problems including real-world implementation of the solution.

To stress the difference between science and engineering, consider the problem: Compute $f(x) = (x^2 - 4)/(x - 2)$ at $x = 2$. In engineering/technology, the answer is 4. This is obtained just by taking the left-hand limit as well as the right-hand limit and observing that these are equal. A simpler numerical way to obtain the value of f(x) at x = 2 in engineering is to compute f(x) at x = 1.99, 2.01. 1.999, 2.001, 1.9999, 2.0001, and observe that these values increasingly become closer to 4. We have assumed in the previous computation sufficiently large, say 14 digit, precision. In fact, the value of f(x) at $x = 2 + 10^{-500}$ as well as at $x = 2 - 10^{-500}$ will each be extremely close to 4. By any measuring/computing device in engineering, we will get f(x) as 4 although exactly at the point x = 2, f(x) is not defined. In science/mathematics, the solution of the problem will be output as undefined (0/0 form).

The function $y(x) = |x|$ is 0 at $x = 0$. The left-hand limit, the right-hand limit, and the value of the function at $x = 0$ are all the same. Hence y(x) is continuous at x = 0. The first derivative of y(x) at x = 0 does not exist as the right-hand derivative

$y'_r(0) = \lim_{h\to 0+} (y(0 + h) - y(0))/h = +1$

while the left-hand derivative

$y'_t(0) = \lim_{h\to 0-} (y(0 + h) - y(0))/h = -1$

and both are different. In engineering/technology, we would say "*y′(0) does not exist*". In science/mathematics, the most precise answer will be "*$y'_r(0)$ exists and is +1 while $y'_t(0)$ exists and is −1 and $y'_r(0) \neq y'_t(0)$*". One might say that this answer implies "the derivative y′(0) does not exist". Strictly speaking, the implication may not tell us the fact that the left-hand derivative does certainly exist as well as the right-hand derivative also does exist. For the sake of preciseness, we, however, still prefer to distinguish these answers.

Consider yet another problem: Compute $g(x) = (\sqrt{(\sin^2 x)})/x$ at x = 0. In engineering/technology, the answer is "*g(0) does not exist at x = 0*". This is obtained by taking the left-hand limit and the right-hand limit and observing that these limist are not equal. One is −1 while the other is +1. A simpler numerical way to obtain the value of g(x) at x = 0 in engineering is to compute g(x) at x = −.001, +.001, −.0001, +.0001, −.00001, +.00001 and observe that these values will alternately tend to −1 and +1. The solution of the problem in science could be output as undefined (0/0 form). However, if we pose the problem as "Compute $g(x) = \lim_{x\to 0} \sqrt{(\sin^2 x)/x}$" then in engineering the answer will be "*the limit does not exist*". In science, the precise answer will be "*the left-hand limit exists and it is −1; the right-hand limit exists and it is +1; both are different*". In fact, the answer in engineering, viz., "the limit does not exist" may not reveal the fact that the left-hand limit exists, so does the right-hand limit. All these are essentially subtle differences. A clear conceptual understanding of these differences does help us in a given context.

From the computation point of view, we will not distinguish between science and engineering computations although we might keep in mind the context while performing computations. However, the precision of computation in science may be significantly more than that in engineering. In fact, in engineering/technology, a relative error (lack of accuracy) less than 0.005% is not, in general, required as it is not implementable in the real world situation and it is hard to find a measuring device which gives accuracy more than 0.005%. We will discuss this accuracy aspect further later in this book.

1.2 Capability and limit of computation

One common feature that pervades both science and engineering is computation. The term computation is used here in the context of a digital computer in a broader sense, viz., in the sense of data/information processing that includes arithmetic and nonarithmetic operations as well as data communication as discussed in Section 1.3. In fact, anything that is done by a computer/computing system is computation. *While mathematical quantities may not satisfy a scientist/an engineer, the numerical quantities do.* A conceptual clarity and quantitative feeling are improved through computation. Till mid-twentieth century, we had computational power next to nothing compared to to-day's (beginning of twenty-first century's) power. To-day tera-flops (10^{12} floating-point operations per second) is a reality and we are talking of peta-flops (10^{15} floating-point operations per second). In fact, the silicon technology on which the digital computers are based is still going unparallely strong. Every 18 months the processing power is doubled, every twelve months the data-communication band-width is doubled while every nine months the disk storage capacity is doubled. The other technologies which might lead to quantum computers or protein-based computers are not only in their infancy but also are not yet commercially promising. These do have some excellent theoretical properties as well as severe bottle-necks.

Capability of computation An important need for computational power is storage/memory. For higher computational power, larger memory is needed since a smaller memory could be a bottle-neck. A rough chart representing storage capacity (bits) versus computational power (bits per second) in both biological computers (living beings including animals) and non-biological (non-living) machines could be as given in Table 1.

Among living computers, the first (topmost) place goes to the whale having a huge memory capacity of 10^{16} bits and a processing speed of 10^{16} bits/sec while among nonliving computers it is the supercomputer (2003) with 10^{14} bits of storage and 10^{13} bits/sec of processing speed in the top position. The British library has 10^{15} bits of information but the processing capability is of order 1, i.e., practically nil. The supercomputing power and storage capacity is dynamic in the sense these are increasing with time while the living computer's power and storage capacity is possibly not that dynamic. It is not seriously possible to distinguish between the nineteenth century human beings and twenty-first century human beings in terms of their memory capability and processing power.

Limit of computation Can we go on doubling the processing power indefinitely? Is there a limit for this power? The answers to these questions are "no" and "yes", respectively. Our demand for higher computational speed as well as storage knows no bound. There are problems, say those in weather forecast, VLSI design, that would take over 1500 hours on today's (2003) supercomputers to be solved. A computer in early 1980s was considered the supermachine if it was capable of executing over 100 million floating point operations per second (> 100 Mflops) with word length of 64 bits and main memory capacity of over 100 million words. Today (2003) it is called a supermachine if it can execute over 1 billion flops (> 1 Gflops) with the same word-length of 64 bits and main memory capacity of over 256 million words. Thus the definition of supercomputers is time-dependent, i.e., yesterday's supercomputers are today's ordinary computers.

Table 1 Memory capacity and computational power of computers

Computers (Living/nonliving)	**Storage capacity (number of bits)**	**Computational power (number of bits/sec)**
Abacus	10^0	10^0
Radio channel	10^0	10^3
Television channel	10^0	10^6
Viral DNA	10^3	10^0
Hand calculator	10^3	10^3
Smart missile	10^3	10^9
Bacterial DNA	10^6	10^0
Bacterial reproduction	10^6	10^3
Personal computer	10^6	10^6
Main frame computer (1980s)	10^8	10^8
Human DNA	10^9	10^0
Honey bee	10^9	10^8
Rat/mouse	10^9	10^{10}
Telephone system	10^{11}	10^{13}
English dictionary	10^{12}	10^0
Video recorder	10^{12}	10^6
Cray supercomputer (1985)	10^{12}	10^{11}
Human visual system	10^{13}	10^{13}
Supercomputer (2003)	10^{14}	10^{13}
Elephant	10^{14}	10^{16}
Human being	10^{14}	10^{13}
British library	10^{15}	10^0
Whale	10^{16}	10^{16}

To discuss about the limit of computation, we should keep the following facts (Alam and Sen 1996) in mind:

1. Classical Von Neumann architecture in which all instructions are executed sequentially has influenced programmers to think sequentially.
2. Programming is affected by both the technology and the architecture which are interrelated.
3. Physics rather than technology and architecture sets up the obstacles (barriers)/ limits to increase the computational power arbitrarily:
   (i) *Speed of light barrier*: Electrical signals (pulses) cannot propagate faster than the speed of light. A random access memory used to 10^9 cycles per second (1 GHtz) will deliver information/data at 0.1 nanosecond (0.1×10^{-9} second) speed if it has a diameter of 3 cm since in 0.1 nanosecond, light travels 3 cm.
   (ii) *Thermal efficiency barrier* The entropy of the system increases whenever there is information processing. Hence the amount of heat that is absorbed is kT $\log_e 2$ per bit, where k is the Boltzmann constant (1.38×10^{-16} erg per degree) and T is the absolute temperature (taken as room temperature, i.e., 300). It is not possible to economize any further on this. If we want to process 10^{30} bits per second, the amount of power that we require is $10^{30} \times 1.38 \times 10^{-16} \times 300 \times 0.6931 / 10^7 = 2.8697 \times 10^9$ watts, where 10^7 erg/sec = 1 watt.
   (iii) *Quantum barrier* Associated with every moving particle is a wave which is quantified such that the energy of one quantum E = hν, where ν = frequency of the wave and h = Plank's constant. The maximum frequency $\nu_{max} = mc^2/h$, where m = mass of the system and c = velocity of light. Thus the frequency band that can be used for signaling is limited to the maximum frequency ν_{max}. From Shannon's information theory, the rate of information (number of information that can be processed per second) cannot exceed ν_{max}. The mass of hydrogen atom is 1.67 $\times 10^{-24}$ gm. c = 3×10^{10} cm/sec, h = 6×10^{-27}. Hence per mass of hydrogen atom, maximum $1.67 \times 10^{-24} \times 3^2 \times 10^{20} / (6 \times 10^{-27}) = 2.5050 \times 10^{23}$ bits/sec can be transmitted. The number of protons in the universe is estimated to be around 10^{73}. Hence if the whole universe is dedicated to information processing, i.e., if all the 10^{73} protons are employed to information processing simultaneously (parallely) then no more than 2.5050×10^{96} bits/sec or 7.8996×10^{103} bits per year can be processed. This is the theoretical limitation (for massively parallel processing) set by the laws of physics and we are nowhere near it! A single

processor (proton) can process only 2.5×10^{23} bits per second or 7.9×10^{30} bits per year and no more!

1.3 What is computation in science and engineering

The word computation conjures up an image of performing arithmetic operations such as add, subtract, multiply, and divide operations on numbers. Undoubtedly these four basic arithmetic operations as well as other operations such as a square-rooting (which involves an infinite sequence of four basic arithmetic operations, in general) do come under computation. Here by computation we would imply a much more general (broader) activity, viz., data (or information or knowledge) processing including data/control signal communication using a digital computer — conventional as well as intelligent. Each and every machine language instruction that a hardware computer executes constitutes a unit of computation. The add, the subtract, the multiply, and the divide instructions, the unconditional and conditional jump/branch instructions, the *for*, *while*, *repeat-until* instructions, *read, write/print* instructions form the building blocks of computation. Each block consists of an ordered sequence of machine instructions. The other instructions such as square-rooting (*sqrt*), sine (*sin*), cosine (*cos*), tangent (*tan*), cotangent (*cot*) computation can be considered higher (macro) level building blocks of computation. One may develop still higher level building blocks such as *inverting a matrix, drawing a least-squares curve* which can be found in MATLAB commands/instructions. Basically, a computer accepts only two distinct symbols, viz., 0 and 1 and operates/manipulates on these symbols in fixed or variable sequences and produces only sequences consisting of these two symbols, viz., 0 and 1 as outputs interpreted according to an appropriate format and context.

Although we talked about computation in a broader sense, *we will limit ourselves with the order of dominant operations for the sake of error and computational complexity*[1].

---

[1]A complex problem or a complex computation implies that the problem/computation is made up of parts; also it implies that it is complicated. The larger the number of parts of the problem/computation is, the more complex it is. For a human being — a living computer — the later implication is often more understood than the former one although psychologically the former implication also would accentuate the difficulty in terms of grasping/remembering. For a nonliving machine — a digital computer — there is absolutely no problem of grasping/remembering or that of ease or difficulty; an analogue of former implication, viz., the number of instructions, measured as amount of computation, is the complexity.

1.4 Tools for Computation

During the pre-computer days, we have been using (i) some kind of writing media such as palm leaves, slates, stone walls/slabs, mud/earth, appropriate materials (plaster of paris), paper and (ii) some kind of writing tools such as ink-pot (ink made of a mixture of powered charcoal and water or liquid colour extracted from plants and/or trees) and pen (pen made of 5 to 6 inches long sharpened bamboo branch or peacock feather or some other stick/brush or some sharp hard object or ball-point pen or fountain pen) combination for doing arithmetic computation as well as graphs, drawings, images/statues — both two dimensional and three dimensional.

During the modern computer days (late twentieth and twenty-first centuries), we use computers as another aid like paper-and-pencil but with much more revolutionary impact on us. If we are asked to compute the positive square-root of a number, say, 30 we could do it over a longer time with or without mistake using paper and pencil provided we know the deterministic arithmetic algorithm for square-rooting. The alternative way is to take out the pocket calculator — programmable or non-programmable — and instantly find the square-root of the number mistakeless (not errorless, in general) by pressing the two number keys 3 and 0 and the square-root operation key. It may be seen that the probability of modern computers committing mistakes in executing an instruction is practically nil while that of any living being — superhuman being or animal or common human being — committing mistake is certainly not nil[2]. However, computers during 1950's and early 1960's did produce wrong output due to circuit failure/malfunction without giving any warning/indication of mistake to us. For example URAL, a Russian computer, that worked with a number system in base 8 during late 1950s/early 1960s did produce occasionally wrong results. A British computer HEC 2M that we had used during late 1950s and early 1960s was relatively good but still not 100% mistakeless. Thus our

---

[2] During one afternoon in early 1970s, we have seen in Indian Institute of Science, Bangalore, India Ms. Sakuntala Devi, a human computer, telling too fast the numerical value of a very long arithmetic expression consisting of terms like $(-72.345)^{6.8911}$ and running from the top corner of a black board to the bottom corner (writing took more than 5 minutes). She took mainly the time to read the expression sequentially and told the answer which was incorrect but within another couple of seconds she told the answer correctly (as computed by the then IBM 360/44 computer! This is not a magic as she has demonstrated this superhuman faculty time and again in various environments/forum. Even a person like her could commit mistakes. How she does this is not known to us nor is it possibly known to her or she is not able to communicate all that goes in her brain.

hypothesis is "*To err is human; not to err is computer (meaning modern computer)*".

The later one i.e., the use of a pocket calculator is so convenient that the present day students in most parts of the world do not at all use the tedious deterministic algorithm for square-rooting. In fact, most of them have forgotten this algorithm and are crippled without the aid of a calculator/computer. During 1940s, 50s, 60s and even 70s, we have been using as input storage media punched cards (standard 80 column IBM cards) and punched paper tapes. For these we have been using enormous natural resources such as trees contributing to eco-hazard to mankind. To-day such media are no longer used — a very positive step and these have become a matter of the past. Even most people of to-day's generation cannot comprehend the extraordinary problems (e.g. card jam at the card reader) associated with these media. Some of them may not even know that there were such cumbersome input storage media.

The present day computer owes its existence to many concepts developed over centuries. The idea of using mechanical aids (such as an Abacus) for calculation is very old. A brief chronological development of digital computers is found in (Alam and Sen 1996). This development may be traced back to the mechanical calculators (1673) due to Pascal and Leibnitz through Babbage's difference engine (1812) and analytical engine (1835) which introduced stored program concept — forerunner of modern digital computers.

A computer may be digital, analog, or hybrid, i.e., a combination of both digital and analog. We will be concerned mainly with the digital computers. A model for computation has two distinct parts, viz., hardware and software. Sometimes we also distinguish another part called firmware (i.e., software implemented in hardware and cannot be easily modified unlike a software) from software. A minimal modern hardware machine must have implicitly or explicitly one central processing unit (CPU) P, one executable (main) memory M, one input-output (IO) processor C, also called channel, one IO device controller K, one input device α_1, say, keyboard, and one output device α_2, say, a printer. If any one of these six units which are appropriately interconnected is absent then the machine cannot be termed as the modern digital computer. Figure 1.1 shows a block diagram of the hardware units of a minimal (simplest) computer.

A generalization of the minimal machine is straightforward. The general computer will have p CPUs P_i, m modules of executable memory M_i, c IO processors C_i, k IO device controllers K_i, and associated with each device controller a number of IO devices α_i, β_i, where each input or each output device could even be another computer. All these hardware units are appropriately connected as shown in Figure 1.2. The dots . . . . . . imply a number of units.

All said and done about the foregoing tools of computation, the most important component present behind these tools for performing computation is the living being/human being. In a sense it is the supreme factor but for which all the foregoing tools would remain useless.

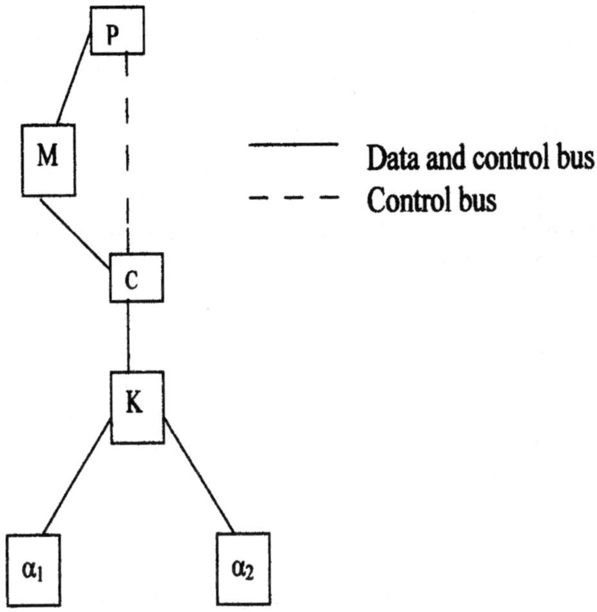

Figure 1.1: A minimal computer

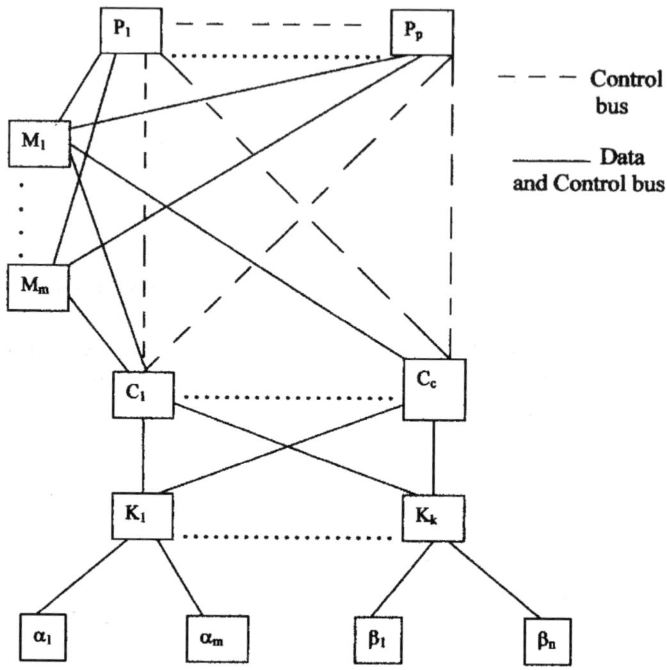

Figure 1.2: A general computer

1.5 Algorithms and Complexity

An algorithm is a finite set of consistent and precise rules, which specifies a sequence of operations to solve all problems of a specific type (completeness). It must have implicit/explicit inputs and explicit outputs. The four basic arithmetic operations and square-rooting are examples of algorithms.

Any computation is based on algorithm(s) — numerical or non-numerical. The word "algorithm" originates from the Arabic word *algorism* meaning the art of computing with *Arabic numerals. The later word is associated with the famous Arabic mathematician Abu-Jafar Muhammad ibn Musa al-Khwarizmi (825 A.D.) who is known to have first suggested the method of adding two decimal numbers by taking one digit from each of the operands and the previous carry digit.*

Algorithms and theorems in mathematics has an interesting analogy. A theorem has axioms and other specified information as inputs and a proof as the output while an algorithm has input data and output results which prove the validity of the algorithm (Figure 1.3).

Input ⟶ **Algorithm** ⟶ **Output (proves the validity of the algorithm)**

Input (axioms and specified information) ⟶ **Theorem** ⟶ **Output (proof)**

Figure 1.3: Analogy between algorithm and theorem

A basic component of every algorithm is iteration (or recursion). Based on the nature of the iteration, i.e., the number of times one or more steps of an algorithm is repeated, algorithms can be classified as *direct, indirect*, and *infinite.* In *direct* algorithms, the number of iterations (repetitions) is known in advance. The conventional matrix multiplication C = AB, where $A = [a_{ij}]$, $B = [b_{ij}]$, and $C = [c_{ij}]$ with $c_{ij} = \sum a_{ik}b_{kj}$ are each an $n \times n$ matrix, for example, is a direct algorithm. The summation is over k = 1(1)n. *Indirect* algorithms contain loops that are repeated an unknown number of times e.g., sieving out primes, partitioning an integer. In an *infinite* algorithm we get better and better estimates of the results the longer we continue the algorithm. The Newton's scheme for finding a root of a nonlinear algebraic or transcendental equation is an infinite algorithm (Krishnamurthy and Sen 2001), where the successive roots will be better and better when the scheme converges.

For solving a problem — a mathematical model — the first step is to get/devise an algorithm, i.e., a finite set of consistent precise rules which can be mechanized. We cannot put every known act in the form of an algorithm. There are problems which are unsolvable, i.e., which cannot be put in an algorithmic form:

(i) *Problem 1* (*Fermat's last theorem*) It is easy to write a program (an ordered set of instructions to a computer) that will search through all positive integers x, y, z, and n (> 2) for a solution of the equation $x^n + y^n = z^n$. Decide whether or not the program will halt if and when a solution is found.

(ii) *Problem 2* Determine, for any programming language, whether or not a given program (a) ever produces an output or (b) can loop for ever on some input or (c) eventually halts on a given input.

(iii) *Problem 3* (*Hilbert's tenth problem*) Devise an algorithm that decides whether or not an arbitrary polynomial with finite number of variables and with integer coefficients has integer zeros.

This algorithmic undecidability/unsolvability is related to the question whether there are statements in a system of axioms that are not provable. Godel (1961) has shown that there are such statements in an axiom system which cannot be proved or disproved (Godel 1962; Nagel and Newman 1964). Analogously, there are tasks for which an algorithmic solution cannot be sought, i.e., these tasks are algorithmically undecidable.

Algorithm as a Turing machine and vice versa Turing observed: "Imagine the operations performed by the computer to be split up into 'simple operations' which are so elementary that it is not easy to imagine them further divided." (Turing 1936; Aho et al. 1974). This observation led Turing to design a machine that is capable of such simple operations. A Turing machine is thus a hypothetical device with an infinite supply of paper tape marked off as square regions. Such a machine is capable of executing only four instructions, viz., (i) move the tape one square left or right, (ii) place a mark on a square, (iii) erase a mark, and (iv) halt. Turing found out that even a machine so simple as this can solve any problem for which an algorithm can be devised/found. In other words, any Turing machine can be expressed as an algorithm and vice versa.

The *computability theory* developed by Turing and other logicians was not concerned with the digital computer resources needed and practicability. Consider the travelling salesman problem (TSP) (Press et al. 1993) — a combinatorial minimization problem:

Find the shortest (least-cost) route which the salesman should take to visit N cities, each only once, with given positions (x_i, y_i) $i = 0(1)N - 1$, returning finally to his city of origin with position (x_0, y_0).

For this problem, the deterministic algorithm has to evaluate $(N - 1)!$ paths (routes) to find the shortest route. The execution time is a function of the input size and this function will be combinatorial which can be shown, *using Sterling's formula* $N! = (\sqrt{(2\pi)}) \times N^{N+0.5} \times \exp(-N + (\theta/(12N)))$, where $N > 0$, $0 < \theta < 1$ (Abramowitz and Stegun 1965), to be exponential (nonpolynomial). It may be seen that $N^{N+0.5} = e^{(N+0.5)\log N}$. Thus this deterministic algorithm is *slow*. No one has, however, discovered a *fast* (execution time is a polynomial of the input size) deterministic algorithm so far. We do, however, have a probabilistic algorithm (based on random choices) called *Simulated Annealling* method (Press et al. 1993; Otten and Ginneken 1989) which is fast (polynomial) and gives an answer which may not be the shortest route although the output is often practically used.

Consider yet another problem of determining a subgraph that is isomorphic to a given graph (Harary 1972). All the known algorithms have execution time that varies exponentially with the number of vertices in the graph and thus are slow. A fast algorithm is yet to be discovered.

Although a problem can be solved in the Turing sense, no efficient (i.e., fast or polynomial-time) algorithm might exist. This fact leads us to the questions: What is algorithmic efficiency? How to measure algorithmic complexity[3], i.e., the amount of computation or the amount of time needed to execute an algorithm? This computational complexity aspect as well as the space complexity (storage needed for the algorithm along with its input data and possibly output result) aspect will pervade the subsequent chapters of this monograph.

It is interesting to note that the simplex algorithm to solve the linear program (LP) *Maximize* $\mathbf{c}^t\mathbf{x}$ subject to $\mathbf{Ax} = \mathbf{b}$, $\mathbf{x} \geq \mathbf{0}$ has been very popular for decades and still is used extensively. This algorithm is mathematically exponential in the worst case although it behaves, for most real-world problems, like a fast[4] (polynomial-time) algorithm. Scientists have been trying to develop a mathematically fast algorithm for decades. The success came only in 1984 with the publication of the projective transformation algorithm by Karmarkar (Karmarkar 1984), which is mathematically fast (polynomial) and has a computational complexity $O(n^{3.5})$, where n is the order of the matrix A in Karmarkar linear program formulation.

Earlier Khachian's ellipsoid algorithm (Khachian 1979) was an interesting development. Although the ellipsoid algorithm is polynomial-time in integer model, Traub and Wozniakowski have shown that it has unbounded complexity in the real number model (Traub and Wozniakowski 1982) discussed later in this chapter. Hence the ellipsoid algorithm is not good in the real number model which is more useful for estimating an algorithm's running (execution) time in actual engineering computation.

Algorithms can also be classified as *deterministic, probabilistic*, and *heuristic*. A *deterministic* algorithm could be direct, indirect, or infinite. All the foregoing examples, viz., the conventional matrix multiplication, sieving out primes, integer partitioning, the Newton scheme are deterministic since we are certain (probability = 1) to get the desired results.

The worst case measure for determining the complexity of an algorithm has been used when we talked about the complexity of simplex and

---

[3] The term "complexity" in computation by an algorithm in the realm of a living computer, say, human being could mean degree of difficulty that is faced by him in grasping/understanding/remembering the algorithm/computation including the amount of computation. Since larger the amount of computation is, more difficult/complicated it is for a man to remember/assimilate. In the realm of nonliving machine, say, a digital computer, such a difficulty does not exist. Here complexity simply implies amount of computation or amount of time needed for execution of an algorithm. The machine does not have absolutely any difficulty in remembering perfectly as it has no life and no feeling. Besides computational/time complexity, we would also talk about space complexity of an algorithm.

[4] All polynomial-time algorithms are called fast algorithms while all exponential (polynomial of degree ∞) algorithms are slow algorithms. To compare two polynomial algorithms A and B where A is $O(n^3)$ and B is $O(n^3/3)$, we prefer to avoid the word *slow* or *slower* since we have already termed exponential algorithms slow.

Karmarkar algorithms. It has been shown that the average case complexity is also important and useful. It is especially so when we deal with algorithms that are based on random choices needing random number generation. Such algorithms are *probabilistic*. The probabilistic methods such as the Monte Carlo methods (Hammersley and Handscomb 1965; Gordon 1970) have been in use several decades. It was shown in 1976 (Rabin 1976) that some problems can be solved more efficiently, i.e., with polynomial-time execution and polynomial-time storage complexity by the probabilistic algorithms than by the known (nonprobabilistic) deterministic algorithms that do not use random choices.

It may be seen that the Monte Carlo method for numerical integration is probabilistic deterministic since by generating more and more uniformly distributed random numbers, each of sufficient length (number of digits), we will obtain better and better integration value, i.e., the value increasingly closer to the exact solution. Here only the numerical error but no mistake exists.

However, Rabin (1976) and Strassen and Solovay (1977) have devised probabilistic algorithms for recognizing prime numbers in polynomial time with a small probability of mistake (rather than error) although basically this problem is in the NP (nondeterministic polynomial)-class. This important finding suggests that the probabilistic algorithms may be useful also for solving other deterministically intractable (difficult to deal with) problems.

An important application of probabilistic algorithms is proving the correctness of a program. The correctness can be shown by constructing suitable "witnesses" for incorrectness using different test inputs. Some randomly chosen test inputs, when such construction is possible, will ensure a provably high probability of correctness. Another application, besides finding the shortest route in a TSP using the simulated annealing (mentioned earlier), is determining a Hamilton path in a graph.

An algorithm could be *heuristic*. A procedure can be sometimes devised to get some guess or intuition about or feel of the problem. Such a procedure is usually termed as *heuristic*, i.e., tending to discover or learn. A heuristic is merely a guide towards finding a solution while an algorithm is a rule for the solution of a problem. A heuristic may or may not succeed in solving a problem. Even if it fails, it may still provide valuable knowledge about how to solve the problem better the next time. A verification procedure may or may not be available in polynomial time to ascertain the correctness of the output result of a heuristic. The polynomial time heuristic algorithm developed by Lakshmikantham et al. (2000) for linear programs has a polynomial time verification while a heuristic algorithm for the chess problem, i.e., the problem of determining the next best possible move in a game of chess does not have a polynomial time verification whether the computed move is truly the best. To find the best possible first move in chess, the machine would have to evaluate 10^{120} game positions. If a machine consisting of all the protons (estimated to be 10^{73}) in the universe could be constructed and run with the speed of light then the current estimate of the age of the universe would be insufficient to find the best starting move using an exhaustive search strategy. Thus a chess-playing program would essentially be a heuristic program. A minimax search is performed to find a good move (not the best one) by terminating it at a fixed depth of, say 4 plies (two moves for each side). If, for example, there is a choice of 30 moves from each position, then we have to evaluate 0.81 million positions for a search of depth 4 (Shannon 1950).

1.6 Types of computation

In fact, all computers essentially perform computation which may be classified as *numerical, semi-numerical, non-numerical,* as well as *symbolic.* The computation will be termed numerical if most of the instructions are arithmetic, i.e., add, subtract, multiply, and divide operations on fixed-point or floating-point numbers. For example, computing a root of a nonlinear equation using the Newton's scheme is numerical. This will be termed semi-numerical if the number of arithmetic instructions (to be executed) and that of non-arithmetic (non-numerical) ones (to be executed) such as branch, loops, read, print are both significant. Generating prime numbers, for instance, involves both numerical as well as non-numerical computations/instructions significantly. The computation is non-numerical if most or all the instructions are non-arithmetic. Searching and sorting are non-numerical. *It can be seen that a meaningful general computer program can be completely non-numerical but not completely numerical.* A symbolic computation is essentially the exact arithmetic (add, subtract, multiply, and divide) computation on symbols and the associated numbers. Finding the determinant of a symbolic square matrix, differentiating/integrating an algebraic or a transcendental function symbolically are examples of symbolic computation. Also, manipulation of strings of symbols such as concatenation, deletion, and insertion is also symbolic computation which may not involve any arithmetic or exact arithmetic operations. This computation does not involve any error, in general. In fact, we do not associate any error to input symbols nor do we generate any error in the output symbolic result. *The symbolic computation is, in general, a combination of errorfree numerical and nonnumerical computation involving symbols.*

1.7 Models of computation

A model of computation (in a restricted sense not involving data communication) has to be specified if we wish to study the difficulty involved in problem solving as well as the error propagation. A model of computation is comprised of the specification of (i) a number system, (ii) arithmetic, and (iii) costs associated with the arithmetic. Examples are

(i) a real number (infinite precision) model of computation that uses real numbers with error-free arithmetic having cost as unit cost per operation,

(ii) a fixed (finite) precision floating-point number model of computation that uses floating-point numbers with floating-point (approximate) arithmetic having cost based on the cost per operation, and

(iii) an integer (fixed point variable precision) model of computation that employs integers with error-free or approximate arithmetic having cost proportional to length of numbers.

Infinite precision (real number) model of computation only exists in nature/material universe. All the computations (processing) strictly following the (consistent) laws of nature are continuously being carried out in a massively parallel fashion in the material universe automatically or by an unseen (by common human being) supreme being. The inputs to all these computations are exact and so are the outputs. Neither the inputs nor the outputs are exactly representable/captured, in general. It is impossible to simulate these computations error-free using any device by human beings. Nevertheless, this model is a very useful mathematical abstraction and helps us to understand the approximate computations better.

The fixed precision floating-point number model is almost universally used for numerical computations whether they occur in science, engineering/technology, or any other area such as economics. Complexity (cost) results are essentially the same as in the infinite precision model. The input quantities are usually erroneous as these are obtained by some measuring device. The outputs are also erroneous; the stability/sensitivity of the algorithm is an important issue here.

The variable precision integer (fixed-point number) model does not model most numerical computations and is used in theoretical studies, e.g., in the complexity derivation of Khachian's ellipsoid algorithms for linear programming (Khachian 1979; Traub and Wozniakowski 1982).

The essential difference between the real number model and the models based on fixed-point/integer as well as floating-point numbers is that *in the former (real number) case the cost of an arithmetic operation is independent of the number (operand) size* while in the later case the cost depends on the operand size. Yet another vital difference is that the real number model is error-free while the other two models are erroneous/approximate, in general.

1.8 Computer representable numbers: Scope and error

Only a very small finite set of rational numbers can be represented in an automatic digital computer since the computer is a finite machine. To do arithmetic in the field of real numbers (R, +, .) using the computer will be unsuccessful since the set of real numbers is infinite and further most of the elements in the set cannot be represented in the computer. This does not imply, however, that the arithmetic cannot be approximated in it.

The approximation of the real numbers R is carried out by using a simulation of these numbers called the *floating-point* numbers F and these numbers are computer representable. Let S be the set of rational numbers. The set F has the following properties.

(i) $F \subset S \subset R$, where the symbol $\subset$ denotes "is a proper subset of". Most of the familiar rational numbers such as 1/3, 5/8, 1/7, 1/10, 2/7 are not elements of F. The set F consists of the rational numbers of the form a/b, where b is an integral power of 2 subject to the precision of the binary computer.

(ii) F is usually symmetric with respect to origin.

(iii) The elements of F are not evenly distributed along the real line. The distance between the two adjacent elements of F near the origin is very small while it becomes increasingly large farther and farther away from the origin.

(iv) The system of floating point numbers (F, +, .) is not a field since closure does not exist under the binary operations mentioned.

Under these circumstances, a logical solution to this problem in many situations is to represent the real number a by its nearest (computer representable) floating-point number $\bar{a}$. thereby introducing the rounding error $e = |a - \bar{a}|$. Rounding errors are further introduced in arithmetic operations due to the lack of closure. If $\bar{a}$ and $\bar{a}'$ are two adjacent elements in F then $a = (\bar{a} + \bar{a}')/2$ is not an element of F. The element a will be $\bar{a}$ or $\bar{a}'$ here.

The effect of rounding errors could be quite serious when we deal with ill-conditioned problems, i.e., the problems in which slight perturbation of the input data would cause very large change in the solution. Consider, for example, the linear system $\mathbf{Ax} = \mathbf{b}$, where A is the coefficient matrix, $\mathbf{x} = [x_1 \; x_2]^t$ is the solution vector to be computed, $\mathbf{b} = [2 \quad 0]^t$ is the right-hand side vector, and t indicates the transpose:

$$64919121x_1 - 159018721x_2 = 2$$

$$41869520.5x_1 - 102558961x_2 = 0$$

The correct solution of this system is $x_1 = 205117922$, $x_2 = 83739041$. The computer outputs, with 14 digit precision and computing $\mathbf{x} = \mathbf{A}^+\mathbf{b}$, the solution vector $x = [x_1 \quad x_2]^t$, where A^+ is the p-inverse (Moore-Penrose inverse) (Lakshmikantham et al. 1996) of the coefficient matrix A, as

$$x = 1.0e-008 * \begin{bmatrix} 0.31081973903851 \\ -0.76134976262938 \end{bmatrix}$$

Observe that

$$Ax = \begin{bmatrix} 1.41247009734806 \\ 0.91097114048343 \end{bmatrix} \neq \begin{bmatrix} 2 \\ 0 \end{bmatrix}$$

and is significantly different from the right-hand side vector $\mathbf{b} = [2 \quad 0]^t$. This is because of ill-conditioning with respect to the inverse of A. The determinant value of the coefficient matrix A is –1 which is small compared to the element size. The condition number defined as $\|A\| \times \|A^{-1}\|$ is too large (inf ($\propto$) produced by MATLAB), where $\| \ \|$ denotes the Euclidean norm.

If we now perturb the matrix A slightly, i.e., if we make

$$A' = A + \begin{bmatrix} 0 & 0 \\ -0.5 & 0 \end{bmatrix}$$

then the determinant of $A' = -79509361$ which is very much different from the determinant of A, viz., –1. The solution vector $\mathbf{x}'$ of the system $A'\mathbf{x}' = \mathbf{b}$ will be, using $\mathbf{x}' = A^+\mathbf{b}$,

$$x' = \begin{bmatrix} 2.57979589914285 \\ 1.05319725299362 \end{bmatrix}.$$

Observe that

$$A'x' = \begin{bmatrix} 2.00000000000000 \\ -0.00000001490116 \end{bmatrix}.$$

which is reasonably close to $\mathbf{b} = [2 \quad 0]^t$.

In the later case, the system $A'\mathbf{x}' = \mathbf{b}$ is well-conditioned with respect to the inverse of A'.

1.9 Problem-solving: Stages and error

Created (by human beings) from the material universe are the physical problems (models). To solve each of these problems, a mathematical model (problem) — simulation/nonsimulation — is derived imposing assumptions and hence errors[5] on the physical model. We translate the mathematical model into an appropriate algorithm (method of solution) and subsequently into a computer program (e.g., MATLAB, high-level Fortran, C, C++, or Java program). Then the digital computer which is always a finite-precision (finite word-length) machine and which can represent only a very small fraction of rational numbers translates this program into its machine language. Finally the computation, i.e., the execution of the machine program takes place and the results are produced.

We like to stress here that the terms *physical problem (model), mathematical model (problem), algorithm, high-level computer program, internal machine representation of this program, and machine program are equivalent* in the sense that each one of these terms is just a set of imperative sentences along with certain given data/inputs/information. These inputs are usually assertive sentences (information).

Consider, for example, the following *physical problem*. 5 liters of milk, 3 liters of sunflower oil, and 12 eggs cost Rs.182. 2 liters of milk, 4 liters of sunflower oil, and 10 eggs cost Rs.190. 7 liters of milk, 5 liters of sunflower oil, and 30 eggs cost Rs. 300. Find the cost of 1 liter of milk, 1 liter of sunflower oil, and 1 egg, where "Rs." denotes "Indian currecncy Rupees".

In this physical model the first three assertive sentences are given data/inputs (information) while the fourth (last) sentence is an imperative one.

The *equivalent mathematical model* is as follows. Given the linear system (i.e., the system of linear equations)

$$\begin{matrix} 5x_1+3x_2+12x_3=182 \\ 2x_1+4x_2+10x_3=190 \\ 7x_1+5x_2+30x_3=300 \end{matrix} \quad \text{or} \quad \overset{A}{\begin{pmatrix} 5 & 3 & 12 \\ 2 & 4 & 10 \\ 7 & 5 & 30 \end{pmatrix}} x = \overset{\mathbf{b}}{\begin{pmatrix} 182 \\ 190 \\ 300 \end{pmatrix}}, \qquad (1.1)$$

---

[5] Error in a quantity is the difference between the exact quantity and the approximate quantity. Since the exact quantity (unless it is measured in terms of number and not amount) is never known and hence the exact error is never known. However, to get at the quality of the result, viz., the quality of the numerical value of the quantity, we would compute an error bound which will be introduced in subsequent chapters.

where $\mathbf{x} = [x_1 \quad x_2 \quad x_3]^t$ is the solution (column) vector, x_1 = cost of one liter of milk, x_2 = cost of one liter of sunflower oil, and x_3 = cost of one egg, compute x_1, x_2, and x_3.

An *equivalent algorithm* is as follows. Given $A\mathbf{x} = \mathbf{b}$, where A is the coefficient matrix of the vector $\mathbf{x}$ in Equation (1.1), $\mathbf{b}$ is the right-hand side column vector in Equation (1.1). Compute $\mathbf{x} = A^{-1}\mathbf{b}$.

An *equivalent computer program* (*high-level*) is a set of input (read) statements/instructions along with an ordered set of instructions (imperative sentences). In a MATLAB program it could be

$A=[5 \quad 3 \quad 12;2 \quad 4 \quad 10;7 \quad 5 \quad 30]$: *Input statement*

$\mathbf{b} = [182 \quad 190 \quad 300]^t$: *Input statement*

$\mathbf{x}$ = inv(A)*$\mathbf{b}$: *Instruction i.e., compute the inverse of the matrix A, post-multiply it by the column vector **b**, and store the result in the column vector **x**.*

The solution $\mathbf{x} = [\,x_1 \ x_2 \ x_3]t = [10 \ 40 \ 1]^t$.

This MATLAB program is translated into a physically larger (in terms of number of instructions) *internal machine representation, say, reverse polish notation* — a lower level program — and then this internal machine representation is translated into the still physically larger *machine language program* (or simply machine program) of the specified computer. The hardware computer understands only its machine program/instructions which are an ordered set of elementary add, subtract, multiply, shift, comparison, test, jump operations and can execute only these machine instructions. There could be several levels of translation/conversion — each is successively physically larger. All of these translated/converted programs in addition to the physical problem, mathematical model, high-level computer program are imperative sentences (simple/compound) with specified data/inputs (assertive sentences). The problem, algorithm, and programs are thus equivalent. However, error, i.e., inaccuracy, is usually injected in each of these stages — problems/programs, in general. In one or more stages in some context, error may not get introduced.

In the execution/computation stage, further error is introduced. Thus the final required solution/result involves the cumulative error. How good this result is has to be established by actually knowing/computing the cumulative error. Diagrammatically, we have (Figure 1.4),

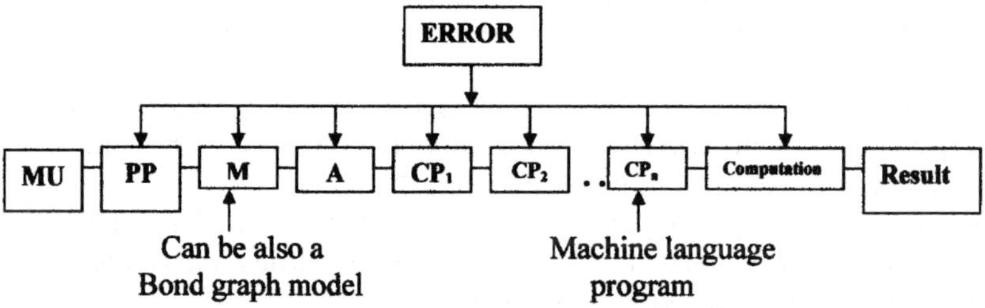

Figure 1.4: Schematic diagram of stages for problem-solving using a digital computer and error injection in various stages

where **MU** = material universe, **PP** = physical problem, **M** = mathematical model, **A** = algorithm, $\mathbf{CP}_1$ = computer program$_1$ (usually a high-level language program, say, MATLAB or Fortran program), $\mathbf{CP}_2$ = computer program$_2$ (could be an internal machine representation, say reverse polish notation — a translated version of $\mathbf{CP}_1$), $\mathbf{CP}_n$ = computer program$_n$ (always a *machine program* of the particular real/physical computer). **Result** is the exact solution (usually numerical) plus the cumulative error — none of these two is ever known in practice. Thus the result obtained out of a computer, though is certainly not the most desired exact solution, is a solution reasonably close to the exact solution, i.e., the error is reasonably low (non-dominant) compared to the magnitude of the solution. It is, therefore, necessary to validate how good the result is. We will discuss this important issue in subsequent sections.

However, it is worth mentioning that for several continuous simulation problems, we may have immediately after the physical problem PP either a mathematical model M or a bond graph model (Cellier 1998) or both in place of M. Bond graphs are a unique way of describing dynamic models. A bond graph is a drawing in which the physical structure as well as the nature of subsystems are shown and can be directly derived from the ideal physical model. When the ideal physical model is partially changed, only the corresponding part of a bond graph has to be changed. For flow problems such as passive and active electric circuits current flow problems, bond graph modelling is a convenient tool. There are softwares that take, as inputs, the bond graphs and produce the required solutions. The physical laws are made use of to derive a bond graph model for a given physical problem. For a bond graph compendium the reader may refer the websites

http://www.eng.gla.ac.uk/bg/

http://www.ece.arizona.edu/~cellier/bg.html

The universe has always an order in it. We have to unearth all the laws governing it (the order). We have unearthed some but many are yet to be unearthed. There is no inconsistency in this order. Observe that *the material universe is absolutely error-free*. Any thing/event that happens in it must follow the laws of nature exactly and, of course, sometimes some of these laws could be beyond our comprehension.

Error-free arithmetic, such as multiple-modulus residue arithmetic, p-adic arithmetic, rational arithmetic (practically not used because of *intermediate number growth*) could be employed only when the inputs are rational (ratio of two finite integers) and the number of operations in the algorithm used is finite (Gregory and Krishnamurthy 1984). For an *ill-conditioned* problem (a problem whose solution produced by using a finite-precision real/floating-point arithmetic has highly pronounced error) involving rational inputs, inexact arithmetic, viz., the real or the complex arithmetic gives totally erroneous results. This fact, usually known as the numerical instability shown in the foregoing section, has been demonstrated by many authors; see, for example, Kulisch and Miranker (1986).

To take care of such an unstable/sensitive situation, the superimposition of an error-free arithmetic on an algorithm is thus not only desirable but also a must in many ill-conditioned problems. Even in error-free implementation (which is possible when the algorithm/method involves only a finite number of arithmetic operations, viz., addition, subtraction, multiplication, and division), the inherent unavoidable error in the input gets magnified in the output results although the computation is 100% error-free. This error could be studied using *interval arithmetic* [Rokne and Lancaster 1971] as well as other methods. This study may be sometimes useful but several times more expensive. Also, the bounds of each exact real quantity may not be reasonably narrow (small) for the computation to be meaningful.

1.10 Stages of problem-solving: Equivalence and hierarchical structure

In Section 1.9 we have seen that we have, starting from the physical problem PP to Computation, n + 4 stages. If n = 3 then we will have 7 stages of problem-solving. The problem is created from the material universe MU and expressed as a PP which may not be corresponding exactly as the problem in MU and in all the n + 4 stages, errors are introduced, sometimes in some stages the error introduced could be zero though. However, an interesting aspect of the step-by-step problem-solving is the equivalence and hierarchical structure of the n + 3 stages (from PP to CP_n both inclusive). These n + 3 stages are equivalent in the sense that each of these stages is a set of imperative sentences along with input data which are usually assertive sentences or information. We measure the size of each stage as the number of imperative sentences/instructions it has. If the PP has k instructions then the mathematical model M has around k instructions. For a balanced hierarchical structure, the algorithm A should have $k_1 = (7 \pm 2)k$ instructions; the high-level programming language CP_1 , say Fortran, will have $k_2 = (7 \pm 2)k_1$ instructions, while the internal machine representation (not visible by the user) CP_2 should have $k_3 = (7 \pm 2)k_2$ instructions and the machine language program CP_3 (when $n = 3$) should have $k_4 = (7 \pm 2)k_3$ instructions. Thus we see that every succeeding stage starting from M to CP_n will be around 5 to 9 times larger physically. The factor 7 ± 2 is introduced since psychologically a common human being can grasp/remember 7 ± 2 things at a time. He can certainly not remember too many items, say 100 names, at a time when told to him once.

This limitation of a common human being, viz., not being able to remember beyond 5 to 9 items at a time, is important in *man-man* (including man-self) communication as well as *man-machine* communication. So far as *machine-machine* communication is concerned, there is no such limitation. The nonliving machine can remember millions of things, say names, for an indefinite period of time exactly once these are given to it subject, however, to its memory capacity. The communication from one machine to another knows no such limitation. As such we may certainly write unbalanced hierarchical structure of a stage (algorithm/program). The machine will face absolutely no problem and it would produce the desired result. Man, however, could have difficulty in grasping as well as in modifying/debugging.

Bibliography

Abramowitz, M.; Stegun, A. (eds.) (1965): *Handbook of Mathematical Functions with Formulas, Graphs, and Mathematical Tables*, Dover Publications, Inc., New York.

Aho, A.V.; Hopcroft, J.E.; Ullman, J.D. (1974): *The Design and Analysis of Computer Algorithms*, Addison-Wesley, Reading, Massachusetts.

Alam, S.S.; Sen, S.K. (1996): Computer and Computing in Fortran 77 (2nd Ed.), Oxford & IBH Publishing Co., New Delhi.

Cellier, F.E. (1998): *Continuous System Simulation*, Springer-Verlag, New York.

Godel. K. (1961): *The Consistency of the Axiom of Choice and of the Generalized Continuum-Hypothesis with the Axioms of Set Theory*, Princeton University Press, Princeton.

Godel, K. (1962): *On Formally Undecidable Propositions of Principia Mathematika and Related Systems,* Oliver and Boyd, Edinburg.

Gregory, R.T.; Krishnamurthy, E.V. (1984): *Methods and Applications of Error-free Computation*, Springer-Verlag, New York.

Gordon, R. (1970): On Monte Carlo algebra, *J. Appl. Prob.* **7**, 373-87.

Hammersley, J.M.; Handscomb, D.C. (1965): *Monte Carlo Methods*, Methuen, London.

Harary, F. (1972): *Graph Theory*, Addison-Wesley, Reading, Massachusetts.

Karmarkar, N. (1984): A new polynomial-time algorithm in linear programming, *Combinatorics*,**4**, 373-395.

Khachian, L.G. (1979): A polynomial algorithm in linear programming, *Dokl. Akad. Nauk USSR*, **244**, 1093-1096, translated as *Soviet Math. Dokl.* **20**, 191-194.

Krishnamurthy, E. V.; Sen, S.K. (2001): *Numerical Algorithms: Computations in Science and Engineering*, Affiliated East West Press, New Delhi.

Kulisch, , U.W.; Miranker, W.L. (1986): The arithmetic of the digital computer: a new approach, *SIAM Review*, **28**, 1-40.

Lakshmikantham, V.; Sen, S.K.; Howell, G. (1996): Vectors versus matrices: p-inversion, cryptographic application, and vector implementation, *Neural, Parallel and Scientific Computations,***4**, 129-140.

Lakshmikantham, V.; Sen, S.K.; Jain, M.K.; Ramful, A. (2000): O(n3) noniterative heuristic algorithm for linear programs with error-free implementation, *Applied Mathematics and Computation* (Elsevier Science Inc., New York), **110**, 2000, 53-81.

Nagel, E.; Newman, J.R. (1964): *Godel's Proof,* New York University Press, New York.

Otten, R.H.J.M.; van Ginneken, L.P.P.P. (1989): *The Annealing Algorithm*, Kluwer, Boston (Has several references to the literature).

Press, W.H.; Teukolsky, A.A.; Vellerling, T.W.; Flannery, B.P. (1993): Numerical Recipe in C: The Art of Scientific Computing (2nd ed.), Cambridge University Press.

Rabin, M.O. (1976): *Probabilistic Algorithms, in Algorithms and Complexity*, ed. J.F. Traub, Academic Press, New York.

Rokne, J.; Lancaster, P. (1971): Complex interval arithmetic, *Comm. ACM*, **14,** 111-112.

Shannon, C.E. (1950): Automatic chess player, *Scientific American*, **182**, 2, 48-51.

Strassen, V.; Solovay, R. (1977): A fast Monte Carlo test for primality, *SIAM J. Comput.*, **6**, 84-85.

Traub, J.F.; Wozniakowski, H. (1982): Complexity of linear programming, *Operations Research Letters*, **1**, No. 1, 59-62.

Turing, A.M. (1936): On computable numbers with an application to the Entscheidungs problem, *Proc. London Math. Soc.*, **42**, (Series 2), 230-65.

Chapter 2

Error: Precisely What, Why, and How

2.1 Introduction

In the universe, the laws of nature are followed perfectly both for matter and for spirit (nonmatter). We consider this statement valid as we have never found out or known any violation of any of the laws of nature. If some laws of nature appear to be violated to somebody, this violation would only imply his imprecise/imperfect knowledge and/or the ignorance of all the concerned laws. We also consider the universe as a gigantic processing/computing device in which massively parallel/concurrent processing is going on non-stop. In fact, this processing is never-ending and possibly never had any beginning. The big bang theory and the steady-state theory are just theories which do not have strict mathematically understood proof or which are based on certain assumptions whose validity is questionable. Further, it would not be wrong to state that all the processing/computations in nature take as inputs exact real quantities and produce as outputs exact results/quantities. These quantities/results can never be representable exactly, in general, by any number system (binary, octal, decimal, hexadecimal, variable radix, negative radix (Krishnamurthy 1971) known to us or would be known in future or by any representation — fixed-point or floating-point — of these number systems. This is because real numbers representing a quantity[1] in nature cannot

[1] According to the Collins Gem English Dictionary, quantity means (i) *specified or definite amount or number*, (ii) *aspect of anything that can be measured, weighed, or counted.* When the quantity is expressed by counting only then it may be exactly represented. However, if the num-

be, in general, represented by *any finite number of digits* in any number system. Hence the exact quantity Q_e and the corresponding represented quantity Q_a are not the same and the difference $Q_e - Q_a$ is the exact[2] error. The exact error thus (as per the definition) can be negative or positive.

The exact error is never known as the exact quantity was not known in the past, is not known now, and will not be known in future. For if it is known then we will not bring the unwanted guest called *error* into picture at all. Strictly speaking, all the quantities that we represent in a computer or in any other form are erroneous in general. It is impossible for us to get rid of error in any computation or in any measurement. A knowledge of error in the inputs as well as that in the output are absolutely necessary if we wish to know the quality of the solution/results (outputs) or if we wish to compare the outputs of two different algorithms for the same problem. Thus, ***error depicts the quality of the results/solution produced by an algorithm.***

2.2 Error: Precisely what and how to compute

2.2.1 What is error

We have seen in Section 2.1 that

(i) the difference between the exact quantity and the approximate quantity is the error,

(ii) error can occur on both the negative side as well as on the positive side of the exact quantity, and

(iii) the exact quantity is never known and hence exact error is never known.

These three facts lead us to modify the definition of error so that this can be used to know the quality of the solution. An obvious way is to define error in terms of an interval in which the exact error lies. Thus **error will always imply *error-bounds*.**

For example, if we read on a resistor its resistance as 500 Ohms with 1% tolerance then the exact resistance lies in [500 – 500 × 1% Ohms, 500 + 500 × 1% Ohms], i.e., between 495 Ohms and 505 Ohms. The absolute error is 5 Ohms or, equivalently, the absolute error bounds are [–5 Ohms, 5 Ohms]. The relative error is 1%, i.e., 0.01 or, equivalently, the relative error bounds are [–0.01, 0.01]. Observe that the relative error/relative error bounds are

---

ber is very large such as 10^{12} (which is the number of neurons in the human brain), 5×10^6 erythrocytes (red blood cells) per mm^3 then the quantity is often not expressible exactly.

[2] The word *exact* in the realm of computation — specifically, numerical computation — implies *error-free*. Thus exact error is nothing but error-free error although we would prefer to use the term *exact error* rather than *error-free error* for easier comprehension.

non-dimensional. If the resistance is 300 Ohms with 5% tolerance then the absolute error is 15 Ohms or, equivalently, the absolute error bounds are [–15 Ohms, 15 Ohms]. These bounds imply that the exact quantity (resistance) lies between 285 Ohms and 315 Ohms. The relative error is 0.05 or, equivalently, the relative error bounds are [–0.05, 0.05]. If the resistance is 2000 Ohms with 5% tolerance then the absolute error is 100 Ohms or, equivalently the absolute error bounds are [–100 Ohms, 100 Ohms]. The exact resistance lies between 1900 Ohms and 2100 Ohms. The relative error is still 0.05 or, equivalently, the relative error bounds are [–0.05, 0.05]. This relative error implies that the exact resistance lies in [2000 – 2000 × 5% Ohms, 2000 + 2000 × 5% Ohms] = [1900 Ohms, 2100 Ohms].

2.2.2 Practically the relative error is all important

In numerical computation, it is the relative error that is/should be almost always used while the absolute error is almost always not used. For instance, 20 million US dollars — is this a big sum or a small sum? This question is clearly meaningless or cannot be answered. Compared to the budget of the United States, this sum may be considered a numerical zero[3] while compared to the monthly salary of an Indian professor, it is very large. Similarly, 10^{-6} — is it a small number or a large number? Once again, it is meaningless or cannot be answered. Compared to 10^{-15}, it is very large while with respect to 1, it is small.

Thus we see that the absolute error is, in practice, useless or has a very limited use although in our subconscious mind, 10^{-6} is considered small and quite often we employ such a figure in a program to come out of a loop. Consequently, such a program could result in a wrong solution for certain problems, i.e., the program is not bug-free and need to be modified replacing the absolute error concept by the relative error concept. The percentage error is, however, just the relative error expressed in terms of percentage. Observe that once the relative error in a specified quantity is known, the absolute error can be readily computed.

Let Q_e = exact quantity, Q_a = approximate quantity. Then the exact error in Q_a could be defined as $E_e = Q_e - Q_a$ or $E_e = Q_a - Q_e$ depending on the sign convention that we may use. Importantly the exact quantity Q_e is never known; for if it is known then we do not have anything to do with error nor do we unnecessarily bring error into the scene. Thus, if we denote error as a nonnegative quantity (conventionally) then we can write $E_a = |Q_e - Q_a|$. This

[3] Specify a small positive number ε. A numerical zero with respect to a given quantity q is defined as any value δ such that $|\delta|/|q| \le \varepsilon$. For example, let $\varepsilon = 0.5 \times 10^{-4}$, δ = \$ 50, q = \$5000000. Then $|\delta|/|q| = 0.1 \times 10^{-4} \le \varepsilon$. Hence δ is a numerical zero with respect to q.

error E_a is called the *absolute error* in the quantity. The *relative error* is then $E_r = |Q_e - Q_a|/|Q_e|$. The *percentage error* is $100E_r$.

2.2.3 A true incident that depicts the importance of relative error

A real-world problem which was being investigated by a research student in 1966 as a part of his research gave rise to a sub-problem of finding the real zeros of a real polynomial p(x) of degree 4. The concerned programmer (not the student) wrote an autocode program (for a polynomial root-finding method) on the then ELLIOT 803 computer and got the zeros. The concerned Ph.D. student substituted one of the four zeros in the polynomial and found the value of the polynomial p(x) as one like 0.434×10^4 instead of zero or a small value, say, of the order of $\pm 10^{-4}$. He also substituted the remaining three zeros one after the other and found that each one of these three zeros produced the polynomial value of the order of $\pm 10^4$. The programmer could not explain to the student why it was happening although the test runs[4] were completely satisfactory. In spite of their best effort, the problem of not getting a small value of the polynomial continued to exist for a couple of days. They then reported the problem to a senior mathematician, who readily told them the solution as follows: A computed real zero r is acceptable if the condition that $p(r + \Delta r)$ and $p(r - \Delta r)$ have opposite signs is satisfied, where Δr is a small real value compared to the value of r. Observe that any polynomial p(x) is a continuous function and if there is a sign change then there has to be a zero. Interestingly each of the four computed zeros did satisfy this condition and the zeros were accepted. The resulting solution of the original problem based on these zeros was satisfactory. This incident might appear trivial. But it might not be so always in real world situation or in our subconscious state of mind. For example, having computed a solution vector $\mathbf{x}$ of a linear system $A\mathbf{x} = \mathbf{b}$, many might like to check by substituting the value (vector) of $\mathbf{x}$ in the expression $A\mathbf{x} - \mathbf{b}$ to see if the value of the expression is $\mathbf{0}$ (null column vector) or close to $\mathbf{0}$. The notation $||A||$ implies the Euclidean norm[5] (Krishnamurthy and Sen 2001) of the matrix A. Unfortunately such a checking could be misleading unless we do error-free computation (in which case, of course, $||A\mathbf{x} - \mathbf{b}||$ will be 0).

---

[4] Test runs are those runs where the zeros of the input polynomials are known a priori.

[5] The Euclidean norm, also called the Erhard-Schmidt norm or the Schur norm or the Frobenius norm or L_2-norm, of a matrix $A = [a_{ij}]$ is defined as $||A|| = \sqrt{}\,(a_{11}^2 + a_{12}^2 + \ldots + a_{1n}^2 + a_{21}^2 + a_{22}^2 + \ldots + a_{2n}^2 + \ldots + a_{m1}^2 + a_{m2}^2 + \ldots + a_{mn}^2)$, i.e., $||A|| = \sqrt{}$ (sum of the squares of all the elements of the matrix A). The Euclidean norm of a vector $\mathbf{x} = [x_j]$ is $||\mathbf{x}|| = \sqrt{}\,(x_1^2 + x_2^2 + \ldots + x_n^2) = \sqrt{}$ (sum of the squares of all the elements of the vector $\mathbf{x}$).

As an illustration of the foregoing polynomial root-finding problem, consider the polynomial equation p(x) = 0, where

$$p(x) = x^3 - 111 \times 10^6x^2 + 111 \times 10^{13}x - 10^{21}.$$

Its three exact roots are 10^6, 10^7, and 10^8. A polynomial root-finding algorithm (which is iterative, in general) produces a root as r1 = 0.99998×10^8 Using the MATLAB commands

```
a=[1 -111*10^6 111*10^13 -10^21]; r1=0.99998*10^8; polyval (a, r1)
```

we obtain p(r1) = −1.781924400800006e+019 which is a value that appears to be far away from exact (ideal) value 0. Yet another root is also found out to be r2 = 1.00001×10^8. Similarly using the MATLAB commands as before, we obtain p(r2) = 8.910189000999961e+018 which also appears to be far away from the exact value 0. Since p(r2) < p(r1) and closer to 0, the root r2 is more accurate. Further since p(r1) and p(r2) are of opposite signs then the exact root r = 10^8 lies in [r1, r2]. The relative error in r1 is given by the MATLAB command

```
er1=abs(r2-r1)/abs(r2);
```

Thus, the relative error $E_{r1} = |r2 - r1|/|r2|$ = 2.999970000299997e-005 is reasonably nice looking although the values of the polynomial at the acceptable roots are not at all good-looking.

2.2.4 Error computation not always given due importance?

While solving a system of partial differential equations — a mathematical model that pervades many areas such as fluid mechanics, weather forecasting, structural engineering — using, say, a finite difference scheme, a rigorous error analysis/computation is often not done. If the numerical solution and/or the corresponding visualization of the solution (through graphs or otherwise) are liked/expected then these are accepted as valid and are reported in journals/proceedings. It may not be rare that the same problem or a variation of it had been solved by other researchers with differing numerical result. Which solution is better? An error computation in both the cases would convincingly/scientifically answer this question. True it is that the error analysis/computation is certainly an additional task needing significant computational resources. But confidence in the quality of the solution is assured scientifically only through error computation. *Sometimes numerical results may be compared with the experimental ones wherever possible.* This comparison may be more rational compared to the subjective feeling/liking of the results.

2.2.5 Definition of computable errors and how to compute them

Since Q_e is never known, the true (exact) absolute error as well as the true (exact) relative/percentage error are never known. Yet we compute some quantity known as the absolute error or the relative error. How do we define these errors? How do we compute them? We answer these questions by denoting

Q = quantity of sufficiently higher order accuracy[6]
or sufficiently more accurate (sma) quantity,
Q′ = quantity of lower order accuracy
or less accurate (la) quantity

and by retaining the foregoing definition of errors, viz. absolute error in Q′ = $E_a = |Q - Q'|$ and relative error in Q′ = $E_r = |Q - Q'|/|Q|$.

Before defining the terms used in the foregoing notation, it is necessary to clearly state the meaning of significant digits in contrast to decimal digits as well as that of the term "correct up to k significant digits".

The ***significance of a quantity*** Q′ is given, in decimal system, as

$$\sigma(Q') = \log_{10}|1/\text{relative error in } Q'|.$$

$\sigma(Q')$ is the number of significant digits up to which the quantity Q′ is correct. On the other hand, $\log_{10}|1/\text{absolute error in } Q'|$ gives the number of decimal digits up to which the quantity Q′ is correct.

When we say that the quantity/result/solution Q′ is correct up to k significant digits, we mean that the relative error in Q′ = $E_r(Q') < 0.5 \times 10^{-k}$. Thus, Q′ is correct at least up to

(i) 1 significant digit if $E_r(Q') < 0.05$, i.e., 5%,
(ii) 2 significant digits if $E_r(Q') < 0.005$, i.e., 0.5%,
(iii) 3 significant digits if $E_r(Q') < 0.0005$, i.e., 0.05%,

---

[6] The word *accuracy* is complementary to the word *error*. That is, if the error is more then the accuracy is less. If we say that no measuring device can give an accuracy of more than 0.005%, it implies that the relative error associated with the device is more than or equal to 0.005%, i.e., 0.00005. Observe that the "accuracy of 0.001%" implies a better accuracy (more accurate) than the "accuracy of 0.005%" while the "relative error of 0.001%" implies less error than the "relative error of 0.005%". However, sometimes, depending on the context, the word "accuracy" has been used to imply "lack of accuracy".or "error". Although such a usage may not confuse in a given context, it is preferable to avoid such a usage.

(iv) 4 significant digits if $E_r(Q') < 0.00005\%$, i.e., 0.005%,

(v) 5 significant digits if $E_r(Q') < 0.000005$, i.e., 0.0005%,

and so on. If $0.005 \le E_r(Q') < 0.05$ then Q′ has exactly 1 significant digit accuracy (not more than 1 and not less than 1). If $0.0005 \le E_r(Q') < 0.005$ then Q′ has exactly 2 significant digit accuracy.

Observe that the relative error is dimensionless. For example, in a container of corn oil, that is supposed to contain 5 liters of oil, we have 4.9 liters of oil. The relative error in the quantity of oil is then 0.02 (not 0.02 liter). In the realm of numerical computation, the term *digit* implies significant digits unless we specifically mention "decimal digits". *The relative error expressed in percent may be called percentage error or relative error or simply error.* There is no confusion in using any of these three terms. For example, if we say that the error in the quantity/result/solution is 5% then this error will imply relative error. If we say that the error is 5% then this implies that the percentage error is 5 (not 5%).

When we say that the quantity Q′ is correct up to k decimal digits (places), we mean that the absolute error in $Q' = E_a(Q') < 0.5 \times 10^{-k}$. Thus Q′ is correct at least up to

(i) 1 decimal place if $E_a(Q') < 0.05$,

(ii) 2 decimal places if $E_a(Q') < 0.005$,

(iii) 3 decimal places if $E_a(Q') < 0.0005$,

(iv) 4 decimal places if $E_a(Q') < 0.00005$,

(v) 5 decimal place if $E_a(Q') < 0.000005$,

and so on. If $0.005 \le E_a(Q') < 0.05$ then Q′ has exactly 1 decimal digit accuracy. If $0.0005 \le E_a(Q') < 0.005$ then Q′ has exactly 2 decimal digit accuracy.

Observe that the absolute error is dimensioned. For example, in a milk packet, that is supposed to contain 1 liter of milk, we have 0.990 liter of milk then the absolute error is 0.01 liter (not 0.01).

Let the precision (word-length) of the computer be sufficiently large compared to the number of digits $k\bar{a}^d$, where $k \ge 1$, d is an integer ≥ 1, and $\bar{a} > 1$. *Let Q′ have an accuracy of some order* $\bar{a} > 1$ *and is correct up to* $k \ge 1$ *significant digits.* Q *will then be a* ***quantity of higher order accuracy or more accurate*** **(ma)** ***quantity*** *if it is correct at least up to* $k\bar{a}^d$ *significant digits* (sufficient condition) and Q′ *will be a* ***quantity of lower order accuracy or less accurate*** **(la)** ***quantity***. If d = 1 then the order of accuracy of Q is $\bar{a}$ higher than that of Q′. If d = 2 then the order of accuracy of Q is $\bar{a}^2$ higher than Q′, and so on. These Q and Q′ are usually known/computed in a fixed-point iteration to obtain the absolute and relative errors. The ***order of convergence*** of a fixed-point iteration scheme will also be referred to as the ***order of accuracy***. We will see that the order of accuracy of the scheme in Example 1 below is 1 while it is 2 in Example 2 (Newton scheme to solve the equation f(x) = 0) below.

We now define the quantity Q of ***sufficiently higher order accuracy*** or ***sufficiently more accurate*** (**sma**) quantity Q as the quantity that satisfies

(i) the ***error-bounds condition***, i.e., the condition that the inequalities $|Q| - E_r|Q| \le |Q_e| \le |Q| + E_r|Q|$, i.e., the exact quantity in magnitude $|Q_e|$ lies in the closed interval[7] $[|Q| - E_r|Q|, |Q| + E_r|Q|]$ and

(ii) the ***more-accuracy condition***, i.e., the condition that Q is closer to Q_e than Q′, i.e., $|Q - Q_e| < |Q' - Q_e|$.

We can certainly compute the foregoing closed interval which is also known as the relative error bounds. But how can we be sure that the exact quantity Q_e lies in this interval? Further how can we be sure that Q is closer to Q_e? To attempt an answer for these questions we will consider a few test[8] examples.

Example 1 The sequence $x_{i+1} = x_i(1 - q) + 1$ $i = 0, 1, \ldots$ till $|x_{i+1} - x_i|/|x_{i+1}| \le 0.5 \times 10^{-4}$ converges linearly (i.e., the order of convergence is 1) to 1/q if $0 < x_0 < 2$ and $0 < q < 1$.

If we take $q = 0.9$, $x_0 = 1.9$ then using the MATLAB commands

```
q = 0.9; x= 1.9;

x = x*(1 – q) + 1
```

where x is taken as x_0, we obtain $x_1 = 1.1900$, $x_2 = 1.1190$, $x_3 = 1.1119$, $x_4 = 1.11119$, $x_5 = 1.111119$ by executing the MATLAB command x = x*(1 – q) + 1 five times. For i = 0, $E_{r0} = |x_1 - x_0|/|x_1| = 0.5966$ is truly a relative error in the quantity x_0 since the exact x, viz., $x_e = 1.11111\ldots1$ lies in the interval $[x_0 - E_{r0}x_0, x_0 + E_{r0}x_0] = [0.7664, 3.0336]$. Thus x_1 in this (first) iteration is an sma solution or a solution of sufficiently higher order accuracy and x_0 is a solution of lower order accuracy. This is, however, not the case for subsequent iterations. For i = 1, $E_{r1} = |x_2 - x_1|/|x_2| = 0.0634$ is not truly a relative error here since x_e does not lie in the interval $[x_1 - E_{r1}x_1, x_1 + E_{r1}x_1] = [1.1145, 1.2655]$. Certainly x_2 is a solution of higher order accuracy (more accurate solution) and x_1 is a solution of lower order accuracy (less accurate

---

[7] This closed interval defines the error-bounds of the exact quantity Q_e. We are 100% confident (unlike the situation in probability and statistics where the confidence is considered always less than 100%) that Q_e is within the interval. This interval should be as small/short as possible when it is produced as the error-bound for the final solution/output. It can be seen that if one wants to be 100% confident that the exact quantity/solution lies in an interval then this interval for a problem in Probability could be too large (or ∞) to be meaningful/useful.

[8] An example/problem is called a test example whose outputs/results/solution are known a priori.

solution) in this (second) iteration but x_2 is not a solution of sufficiently higher order accuracy. Similarly, $E_{r2} = |x_3 - x_2|/|x_3| = 0.0064$ is also not truly a relative error since x_e does not lie in the interval $[x_2 - E_{r2}x_2, x_2 + E_{r2}x_2] = [1.1119, 1.1261]$. x_3 is certainly a solution of higher order accuracy and x_2 is a quantity of lower order accuracy in this third iteration but x_3 is not a solution of sufficiently higher order accuracy.

Although the sequence converges to a solution, we cannot say with 100% confidence that the number of significant digits up to which the solution is correct is 4 from the stopping condition (i.e., the *relative error condition*), viz., $|x_{i+1} - x_i|/|x_{i+1}| \leq 0.5 \times 10^{-4}$.

The error bounds in this example do not contain the exact solution although in most numerical computation, we obtain error bounds which do contain (bracket) the exact solution; in fact, we are 100% confident about localizing the exact solution within the bounds. ***We do not bring in or state the confidence level explicitly in deterministic/nonprobabilistic numerical computations in general; implicitly we take this level as 100% to specify the error bounds quite unlike the statistical/probabilistic computations.***

In a fixed-point iteration scheme (Krishnamurthy and Sen 2001), if the order of convergence of the scheme is greater than 1 then the concerned successive relative error bounds would possibly encompass the exact solution subject to the precision of the computer. A mathematical study along with numerical experiments on the order of convergence and corresponding error bounds would make us 100% confident about the ***correctness*** of the error bounds, i.e., whether the bounds really bracket the exact solution.

Example 2 Now let us consider the Newton scheme (Krishnamurthy and Sen 2001) to obtain a root of the nonlinear equation $f(x) = 0$, where $f(x)$ is continuous and differentiable. The scheme is, for 4 significant digit accuracy,

$$x_{i+1} = x_i - (f'(x_i)/f(x_i)),\ \ i = 0, 1, 2, \ldots,\ \text{till}\ |x_{i+1} - x_i|/|x_{i+1}| \leq 0.5 \times 10^{-4},$$

where x_0 is an initial approximation of the root (to be specified by the user) and $f'(x) = df/dx$. The sequence x_{i+1} $i = 0, 1, 2, \ldots$, has an order of convergence 2 (hence the order of accuracy 2) and converges to a root of the equation $f(x) = 0$ when it converges. For polynomials, the scheme converges even if the initial approximation x_0 is far away from a true root.

To find the square-root of a given number y using the Newton scheme, we take $f(x) = x^2 - y = 0$. Hence the sequence

$$x_{i+1} = (x_i + (y/x_i))/2 \quad i = 0, 1, 2, \ldots,\ \text{till}\ |x_{i+1} - x_i|/|x_{i+1}| \leq 0.5 \times 10^{-4}$$

will always converge for any finite initial approximation x_0 assuming a sufficiently large precision of the computer. If $y = 25$ and $x_0 = 500$ — a value far away from the exact (nearer) root $x_e = 5$ — then $x_1 = 250.0250$, $E_{r0} = |x_1 - x_0|/|x_1| = 0.9998$. The exact root, viz., x_e lies in $[x_0 - E_{r0}x_0, x_0 + E_{r0}x_0]$ = [0.1000, 999.9]. Thus, x_1 is an sma solution while x_0 is a solution of lower order accuracy although x_1 is still far away from x_e. $x_2 = 125.0625$. $E_{r1} = |x_2 - x_1|/|x_2| = 0.9992$. The exact root, viz., x_e lies in $[x_1 - E_{r1}x_1, x_1 + E_{r1}x_1]$ = [0.2000, 499.8500]. $x_3 = 62.6312$, $x_4 = 31.5152$, $x_5 = 16.1542$, $x_6 = 8.8509$, $x_7 = 5.8377$, $x_8 = 5.0601$, $x_9 = 5.0004$, $x_{10} = 5.0000$. The Newton scheme (order of convergence is 2) always satisfies the condition for sufficiently more (sm) accuracy, viz., sufficiently higher-order accuracy for well-conditioned polynomials (i.e., polynomials whose zeros[9] are not too closely spaced) with distinct zeros even with a bad initial approximation. It may be seen that for multiple zeros, the Newton scheme enters into oscillation of the iterates x_i around the zero. The derivative $f'(x)$ tends to 0 faster than $f(x)$ in the case of multiple zeros and hence oscillation results as we always work with a finite precision. The deflated Newton scheme is the remedy for such an oscillation (Krishnamurthy and Sen 2001).

After a successful completion of the foregoing Newton scheme, we are 100% confident that the computed root is correct at least up to 4 significant digits. This confidence assumes that the input data are exact and the precision of the machine is sufficiently large.

For noniterative algorithms, we have to have the knowledge of the sma quantity/solution along with the la quantity/solution so that we can say about the quality of the solution, i.e., how many significant digits the solution is correct.

How do we know that the solution is sma How do we ascertain that a solution is an sma solution or simply an ma solution? To ascertain, we need the knowledge of la solution as well as some mechanism, e.g., changing some parameters, computing the next iteration solution, to produce a solution through the algorithm and comparing this solution with the la solution. This depends on the specified problem and the concerned algorithm. Sometimes laboratory/field experiment or numerical experiment could be helpful. No general guidelines can be stated to answer this question (independent of algorithms/problems). We will discuss this issue when we deal with error for the specified problem/algorithm in subsequent chapters.

2.2.6 Problems in nature/universe

Since the dawn of civilization, man is trying to understand the nature to make the best use of natural resources and laws of nature for his own com-

---

[9] Zeros of a polynomial $f(x)$ are exactly the same as the roots of the polynomial equation $f(x) = 0$.

fort and benefit. This attempt has given rise to numerous engineering/scientific problems which need to be solved. Constructing a bridge over a river, building a supersonic jet aircraft or a spacecraft, designing and developing a robot that could search a sea-bed for retrieving materials/body-parts of an aircraft that met with an accident over a sea, forecasting weather are a few of the problems. In order to know the quality of the numerical solution associated with these problems, error bounds should be computed to validate the solution or to know the quality of the solution.

2.2.7 Ever-existing error in measuring devices

Associated with any measuring device there is a fixed order of error The problems in nature/universe cannot be often written/expressed exactly unless we deal with discrete countable objects. The mangoes in a basket are countable and will have exact representation so far as their numbers are concerned. If we deal with the weight of the mangoes then this weight is neither countable nor can this be expressed exactly since associated with any *measuring instrument* there is an *order of error*. A screw gauge that is used to measure the diameter of a sphere of size, say, 4 cm (within the range 1 cm – 7 cm) may have an error of the order of 6×10^{-4} cm (i.e., the exact diameter lies in [$4-6\times10^{-4}$ cm, $4+6\times10^{-4}$ cm]). A measuring tape that is used to measure the length of 150 meter (within the range 50 meter – 200 meter) of a play ground may have an error of the order of, say, 20 cm (i.e., the exact length of the play ground lies in [(15000 – 20) cm, (15000 + 20) cm]). A weighing machine that is used to measure gold in the range 5 gm – 20 gm may have an error of the order of 15 mgm (i.e., 15×10^{-3} gm) while weighing platform/bridge that is used to measure a loaded lorry of say 5 tones may have an error of the order of 10 kg. Thus, ***associated with any measuring instrument, there is a fixed order of error and this order varies from one measuring instrument to another***. Further, ***almost all measuring instruments will have an error (relative) not less than 0.005 % (i.e., 0.5×10^{-4})***. This implies that it is useless in practice to produce final numerical results (through computation) with a relative error less than 0.5×10^{-4}. Thus most of the time for the final numerical (iterative) solution, we can introduce test (in a computer program) whether the relative error in a quantity falls below 0.5×10^{-4} or not. It is not necessary in practice for the quantity to have its relative error, say 0.5×10^{-8} (i.e., less than 0.5×10^{-4}) as it will not serve any purpose in any real world situation/implementation. However, in the intermediate steps, higher-order accuracy would often be required so that the final result that will be used for actual engineering implementation has error (i.e., order of error) 0.5×10^{-4}. To achieve a relative error less than 0.5×10^{-4} will have no other negative effect except the extra computing cost subject, however, to the precision (word-length) of the computer used. *In the foregoing computation, we have assumed that the input data are error-free.*

2.2.7.1 Order of error: absolute versus relative

Depending on the context the order of error associated with a measure will imply absolute error (i.e., absolute error bounds) or relative error (i.e., relative error bounds). When it implies absolute error bounds, it is ***fixed*** for a measure and does not change when the measure is used to measure different quantities. When it implies relative error bounds, it is ***variable*** for a measure and does change when the measure is used to measure different quantities. The relative order of error (of a measure) changing from one quantity to another different quantity can be seen from Section 2.2.7.

2.2.8 Injection of error by measuring device and assumption

The problems in nature/universe are errorless but as soon as we, the human beings, write/specify the equivalent physical problems, error will be automatically injected into these physical problems due to (i) the *nonexactness* of the reading of real quantities by a measuring instrument and (ii) the *assumptions* (if any) made to permit a solution relatively easily. Observe that the motive of the assumptions is essentially to make a solution of the given problem possible/less complex.

Consider, for example, the *prey-predator* problem. Let x(t), y(t) be the populations of the prey and predator species at time t. We assume that (i) if there are no predators, the prey species will grow at a rate proportional to the population of the prey species, (ii) if there are no prey, the predator species will decline at a rate proportional to the population of the predator species, and (iii) the presence of both predators and preys is beneficial to the growth of predator species and is harmful to the growth of prey species – specifically the predator species increases and the prey species decreases at rates proportional to the product of the two populations.

These three assumptions are used to obtain the physical problem which does not exactly present the corresponding problem in nature/universe and which can be more easily solved. The actual problem in the nature/universe cannot often be so easily and exactly written as a physical problem because there could be many more parameters such as the climatic condition, pollution, natural disaster including earthquake, hurricane/cyclone. Even if we are able to write the physical problem exactly, the solution of this exact problem could be too difficult or not possible. *The assumptions though inject into the physical problem error help us making the problem relatively simple and more easily solvable.*

Then comes the following mathematical model which is a system of nonlinear first-order ordinary differential equations (ODEs).

Compute x(t), y(t), for different values of t, from the ODEs

$$dx/dt = ax - bxy, \quad dy/dt = -py - qxy \ , \ a, b, p, q > 0 \qquad (2.1)$$

where, at $t = 0$, $x = x_0$, $y = y_0$ (initial conditions).

The algorithm, programs, and computation follow. Error may be injected in each of these stages. The output/result then has the cumulative error embedded in it. Observe that the concerned problem in nature/universe will produce result (viz., the prey and predator species at a specified time) completely error-free. Maybe we, the human beings, cannot exactly specify the problem nor can we get the result in exactly the way nature gets it. Thus **nature is the best modeler and the best (infinite-precision) computer that obtains the real result exactly without any trace of error anywhere. It does not make any assumption nor does it need any measure.**

2.2.9 Relative limitation/incapability of measure

We will define the absolute limitation of a measuring device as the incapability of the device to measure exactly a given real quantity q. The absolute error $|q - q'|$, where q′ is the value produced by the device, gives us the *extent* of incapability/limitation. Let this quantity be the electric current flowing through a given copper wire. Let the exact value of the root-mean square current (not known) be a amp. If the device shows the reading as a′ amp then the absolute error bounds associated with the device for this current are $[-|a - a'|$ amp, $+|a - a'|$ amp$]$. These bounds define the extent of absolute limitation of the device for the current under consideration. Any of the infinite possible currents that lie between a amp and a′ amp, both inclusive, will not be detected as one different from a′ amp. There is no way to know this exact current a which is a real quantity and hence there is no way to know the extent of exact absolute limitation of the device with respect to a given quantity. The absolute limitation will be 0 or, equivalently nonexistent if the device is capable of measuring the quantity exactly. This could be only possible when the quantity is measured in terms of numbers (not in terms of weight or volume or length or time). To know an approximate extent of the absolute limitation in a less accurate (la) device, one could use a sufficiently more accurate (sma) device if it is available. Observe that if an sma device is available then there may be no justification of bringing an la device into picture. In the absolute limitation, two quantities, say, two distinct quantities of current flowing through two different wires are not compared. In other words, we have *only one quantity and only one measuring device when we talk about absolute incapability of a device. To know the extent of absolute incapability, we need to have another sma device equivalent*[10] *(capable of measuring large quantity) to la device.*

In contrast, *in the relative limitation/incapability of a device*, we have *two (or more) nearly equal quantities that are compared and only one measuring device. To know the relative ordering of the quantities, we need another device capable of measuring very small quantities (viz., the difference) independent of the large quantities.* For example, if a device is used to measure two distinct but nearly equal quantities and if it is not able to distinguish the difference between these quantities to say which is smaller then we will call this incapability of the device as its relative limitation. We may not be interested to know the extent of relative incapability. We will not be able to know this extent until we have a means or device capable of determining the difference (very small) independent of the two actual quantities (very large). However, we would be interested to know the order of the two (or more) nearly equal quantities to ascertain which is smaller (or smallest). Consider a few examples to illustrate the relative incapability of measuring devices.

Weights of live and dead bodies: Different? Can we measure the weight of a human being just before and just after his death? The dying person could be on a light bed rested on a high-precision weighing platform. The weights can be electronically recorded every second or every few seconds along with the record of other physiological parameters which decide the clinical[11] death of a person under controlled condition.

In one very limited 1907 experiment (Ogden 2000), researcher Duncan McDougall attempted to measure the difference between the weight of a live body and that of the body immediately after death by weighing five patients as they died. In two patients, there was a sudden weight loss of half an ounce, followed by another sudden one-ounce weight loss within three minutes of the time of death. A third patient's weight fluctuated just after death, first dropping a bit, followed by an abrupt large weight gain (!), then weight loss. There was no discernable change on the other two patients. The results thus were inconclusive. It is not difficult to imagine that nearly 100 years ago the scientists did not have a high precision weighing platform as we have to-day. When we are attempting to measure the weight of two nearly equal bodies where the difference is a very small fraction of the body weight, the measuring device could fail to detect which body is heavier. Further any

---

[10] Equivalent in the sense that the la as well as sma devices measure the same (large) quantity.

[11] Clinical death is that death declared according to the norms of medical science. This death may not be always the true death. A true death is one from which the body does not certainly return to life. There seems to be no very sharp demarcation between these two deaths. Certainly one can ascertain that the true death has taken place when the body has decomposed to such an extent that the process is irreversible.

statistical experiment needs reasonably large set of data (i.e., a large number of dying patients) before we could come to an emphatic conclusion assuming a high precision of the scale used.

True it is that the weight of the body does change at least due to the breathing in and breathing out process, although such a change cannot be measured since the weight of the oxygen intake is a numerical zero compared to the body weight.

Assume that the weight of the live body just before death is 60 kg. Just after death if it is 60 kg – 0.01 oz where 1 oz = 0.0283459 kg then this implies that the order of error of the weighing platform (machine) is $(0.0283495 \times 0.01 \times 100)/60\% = 4.7249167 \times 10^{-4}\% = 0.00047249\%$ for the body. Is the machine thus too accurate (error is much less than 0.005%)!! This accuracy of the weighing platform is certainly questionable if we attempt to know the difference in weight using the foregoing weighing platform.

However, if we neglect the weight fluctuation due to breathing (oxygen mixed with nitrogen) we are not sure that there is such a difference between the weight of a live body and that of the body immediately after death. This difference might be there or might not be there. We are yet to invent a measuring device which would significantly be more accurate than 0.005% and would be able to answer these questions and several other questions in the realm of matter and nonmatter. In fact, are we able to determine the exact point of time before which the body was live and after which the body was clinically dead (and vice versa if the true death has not occurred)?

Weights of live and nonlive wires: different? Is there a difference in weight between a live (electric) wire and the same nonlive wire? Certainly there is the most important difference between these two wires in terms of the physical parameter, viz., the current — one carrying electric current and the other carrying practically no current. Is there a difference in terms of elementary/fundamental particles (which are matter)? Yet we are still not sure if weights differ. One might have firm conviction that weights are absolutely identical but he cannot prove this conviction beyond doubt possibly due to the relative limitation of the measuring instrument. Any measuring device does not permit too huge a range of measurement (e.g., 10^{-6} gm to 10^{6} gm, 10^{-6} amp to 10^{6} amp).

Weights in normal and samadhi states: different? In *Nirvikalpa Samadhi* (Satyananda 2000; Prabhavananda and Isherwood 2002) the person does not breathe (no oxygen intake), heart does not function (no pumping of blood), blood does not flow, the body becomes cold like a dead body, thought ceases to exist. Any medical examination will declare such a person with these physiological changes dead. However, he comes back to life by *presuggestion* or bringing a thought into the body system. If this thought — say, thought of drinking water — self-suggested before going to *samadhi* is not brought into the body system, the person will possibly never return to life and his body will perish in course of time. The measurement of the foregoing physiological parameters pulse, heart beats, blood pressure, body temperature, brain function are possible (though not very accurately). All these measurements fail to tell us whether the man is in *samadhi* or not. Is there any weight difference between normal state and *Nirvikalpa* state? We are not sure. Nobody can be emphatic that the difference does not exist since he does not have a measuring device that has truly very high precision (say, accuracy of the order of $0.5 \times 10^{-26}\%$). Further, coming across persons in *samadhi* state may be quite difficult. Consequently, a statistical experiment cannot be carried out.

Even an electrical/heat conductor weighing, say, 0.5 kg loosing a few electrons from its surface due to some reason will not depict any weight change in any existing measuring device although the weight of the conductor before loosing electrons and that after loosing electrons are certainly different. An electron is a fundamental particle and has 9.1095×10^{-31} kg as its (approximate) weight.

2.2.9.1 How to know difference between two almost identical quantities

We have already seen that a tool for measuring very large quantities cannot measure very small quantities at all or reasonably accurately. A measure for very small quantities cannot be even used to measure very large quantities.

In the foregoing examples, if we are able to know how many electrons escaped, how much oxygen absorbed at a given instant, then we will be able to know the weight of these electrons or the weight of oxygen at that instant assuming a controlled environment where all the other parameters are unchanged. This weight is not relative to the weight of the body. Consequently, we would be possibly able to know the difference in weight of a live body and that of a dead body at that point of time. ***We, therefore, need to use two different devices — one for measuring very large quantity and the other for very small quantity to decide which quantity is smaller — when we have to order two nearly equal quantities.*** The measuring instruments used for measuring the weight of a body which is much much higher than these weights would never be able to record the difference. Hence we should ***avoid measuring two nearly equal (e.g., by weight or by volume) quantities where the relative difference is too small, say less than 0.005% just by using one measure for large quantities.***

Where time measurement does not help, television pictures help Consider a 100 meter sprint in which two participants start sprinting at the same time. If the time of the sprint, measured electronically correct up to 2 decimal places in seconds, happen to be the same, say 9.81 sec, for two participants then we would possibly declare them joint winners (such a declaration has never happened so far in any olympic or world athletic meet) if there is no other way to rank. The ***television pictures*** of the finish from different angles could possibly help to rank the sprinters decisively. In fact, in cricket which is currently one of the most intoxicating games in the Indian sub-continent and also played in Australia, New Zealand, England, South Africa, the West Indies, Kenya, Namibia, Canada, Holland, and a few other countries, such television pictures decide run-outs quite satisfactorily. A run-out is defined as the situation in which the cricket ball hits the wicket before the batsman could reach the crease. The electronic measurement of time, when possible might not help a run-out decision. During 1950s and 1960s, such facilities were not existing or not used in cricket. The umpire's decision (that might have human error) was accepted as final sometimes with dissatisfaction of one of the two teams.

Where television pictures do not help, sound helps In some instances in cricket where human vision or even television pictures do not conclusively decide catch-outs, the ***sound*** amplified by an amplifier kept concealed near/at the wicket could decide catch-outs satisfactorily. A catch-out is defined as the situation where the ball touches/hits the bat and finds its place in a fielders hand without touching the ground.

2.2.10 Measures: Capabilities/limitations with more examples

Consider a weighing bridge that is capable of measuring weight up to 15 tones. Let a loaded truck weigh 13.89 tones on the bridge. If we take out one kilogram from the loaded truck or add one kilogram to it, will the bridge be able to detect the difference as approximately 1 kg? The answer can be easily seen to be "no" when one comes across such a weighing bridge.

When we wish to measure the distance of the sun from the earth at a given time, we might send a beam of monochromatic light (say, a laser beam) and measure the time for the light to go to the sun and to come back after being reflected. Knowing the time, say 16.67 minutes and the velocity of the foregoing light, say $v = 186000$ miles per sec, we can get the distance as $d = v \times 60 \times 16.67/2 = 93018600$ miles. If we repeat the experiment of measuring under the real dynamic conditions, e.g., a condition that the earth is rotating around its own axis with the speed of 18 miles per sec then we may not get the same foregoing d. The absolute error could be 4600 miles or more.

When we measure the wavelength of a monochromatic light, say a sodium light, in terms of angstrom (10^{-8} cm) using an optical means, we will have error which is not usually less than the relative (percentage) error 0.005%. When we measure the speed of a cricket ball electronically in terms of km/hr, the error involved is not usually less that 0.005%. *It would thus not be wrong to say that associated with almost every instrument (electronic or not, optical or not, sound or not) measuring weight, length/height/width/depth, or time, there is an error which is greater than or equal to 0.005%, however much accurate the instrument is.*

2.2.11 Error in solution which is vector/matrix: Need for norm

A solution/result need not be scalar. It could be vector or a matrix. How do we specify error in the solution? To answer this question we consider the consistent linear system $A\mathbf{x} = \mathbf{b}$, where $A = [a_{ij}]$ is an $m \times n$ numerically known matrix, $\mathbf{b} = [b_j]$ is a numerically known column vector of dimension m. The problem is to find a value (vector) of the vector $\mathbf{x}$ and a relative error in $\mathbf{x}$. Let A^+ be the minimum norm least squares inverse (also known as the Moore-Penrose inverse or the pseudo-inverse or the p-inverse (Lakshmikantham et al. 1996)) of the matrix A then the general solution can be written as $\mathbf{x} = A^+\mathbf{b} + (I - A^+A)\mathbf{z}$, where I = the $n \times n$ unit matrix and $\mathbf{z}$ is an arbitrary column vector of dimension n. This general solution will be a true solution if $A\mathbf{x} = \mathbf{b}$ is consistent else this will be the minimum norm least squares solution when $\mathbf{z}$ is taken $\mathbf{0}$ (null column vector). Yet another problem is to compute the inverse A^+ and the relative error in it. *The inverse A^+ will be the true inverse A^{-1} if the matrix A is nonsingular, i.e., A is square and its determinant is not 0.* The matrix A^+ satisfies the four conditions $AA^+A = A$, $A^+AA^+ = A^+$, $(AA^+)^t = AA^+$, and $(A^+A)^t = A^+A$, is always unique, and always exists for any real or complex $m \times n$ matrix A.

The equations $A\mathbf{x} = \mathbf{b}$ will have either no solution (contradictory/ inconsistent equations) or just one solution or infinite solutions. These cannot have just two or just three or just k (k is any finite positive integer) solutions. For, if these have then a linear combination of these finite number of solutions is also a solution implying the existence of infinity of solutions. For example, the system of equations $3x_1 + 4x_2 = 7$, $6x_1 + 8x_2 = 13$ has no solution, i.e., we will never be able to find a numerical value of x_1 and that of an x_2 which will satisfy both the equations simultaneously. Geometrically, these two equations will represent two one-dimensional hyperplanes which are here straight lines and which are non-coincident parallel lines, i.e., these two lines will never intersect implying no solution. If 13 is replaced by 14 in the foregoing equation then we will have infinite solutions – one solution is $x_1 = 1$, $x_2 = 1$ while another solution is $x_1 = -1/3$, $x_2 = 2$. Geometrically, the later two equations represent two coincident 1-dimensional hyperplanes implying infinity of points of intersections, i.e., infinity of solutions. If we have the equations $3x_1 + 4x_2 = 7$ and $6x_1 + 7x_2 = 13$ then there is only one solution, viz., $x_1 = 1$, $x_2 = 1$. Geometrically, these two equations will represent two non-parallel straight lines that intersect at the point (1,1). Here we will compute a solution along with an error as well as the inverse A^+ also with an error.

The matrix inverse has $n \times m$ elements and the solution vector has n elements. Associated with each element there is an error. Are we then going to compute $n \times m$ errors as well as n errors, respectively? The answer is certainly 'no'. From ***human psychology*** point of view, we are able to compare two values at a time and can say one is larger than the other. If there are two different methods/algorithms to compute A^+ and if we compute $n \times m$ errors corresponding to $n \times m$ elements of A^+ for each method then we will not be able to compare and say which method has given better result, i.e., has produced less errors unless we introduce some way to produce a single figure/value for errors for each method and then compare these two values to conclude logically that one is better than the other. A way to produce a single figure is to compute a norm of a matrix or a vector. There are different kinds of norms – Euclidean norm, spectral norm, maximum row-sum(L_∞-) norm, maximum column-sum (L_1-) norm (Krishnamurthy and Sen 2001). We will restrict ourselves to Euclidean norm. The Euclidean norm for an $m \times n$ matrix A is defined, using the summation notation, by the real non-negative number $||A|| = (\Sigma\Sigma\ |a_{ij}|^2)^{1/2}$, where the first summation will run over $i=1$ to m and the second summation from $j=1$ to n. Similarly the Euclidean norm for an n-dimensional vector $\mathbf{x} = [x_j]$ is $(\Sigma|x_j|^2)^{1/2}$, where the summation will run from $j=1$ to $j=n$.

A simple iterative method (Sen and Prabhu 1976, Krishnamurthy and Sen 2001, Sen 2002) with quadratic convergence to compute the minimum norm least squares inverse A^+ for a numerically specified $m \times n$ matrix A is as follows (denoting by the superscript t the transpose, by tr the trace, and by I the unit matrix of order m).

Define $X_0 = A^t / tr(AA^t)$.
Compute $X_{k+1} = X_k(2I - AX_k)$,
$k = 0, 1, 2, \ldots$, till $||X_{k+1} - X_k||/||X_{k+1}|| \le 0.5 \times 10^{-4}$.

The matrix X_{k+1} will be the required A^+ correct up to 4 significant digits after the execution of the foregoing method.

Let us compute the minimum norm least squares inverse for the given matrix

$$A = \begin{bmatrix} 3 & 2 & 1 \\ 1 & 1 & -1 \end{bmatrix}$$

Here I is the 2 × 2 unit matrix. The trace $(AA^t) = 17$.

$$X_0 = \begin{bmatrix} .1765 & .0588 \\ .1176 & .0588 \\ .0588 & -.0588 \end{bmatrix},$$

$$X_1 = X_0(2I\text{-}AX_0) = \begin{bmatrix} .1938 & .0657 \\ .1246 & .0796 \\ .0830 & -.1211 \end{bmatrix}, \quad \|X_1 - X_0\|/\|X_1\| = .2728 > .5\times10^{-4},$$

$$X_2 = X_1(2I\text{-}AX_1) = \begin{bmatrix} .1951 & .0713 \\ .1030 & .1500 \\ .1705 & -.3886 \end{bmatrix}, \quad \|X_2 - X_1\|/\|X_2\| = .4475 > .5\times10^{-4}.$$

The successive norms $||X_{k+1} - X_k||/||X_{k+1}||$ for $k = 2, 3, 4, 5, 6, 7$ are .3962, .3010, .1564, .0332, .0012, and 1.3987×10^{-6}, where the last norm satisfies the condition, viz., $1.3986 \times 10^{-6} < .5 \times 10^{-4}$. Therefore,

$$X_8 = A^+ = \begin{bmatrix} .1923 & .0769 \\ .0769 & .2308 \\ .2692 & -.6923 \end{bmatrix}$$

is the required minimum norm least squares inverse correct up to 4 significant digits. Thus the relative error in each element is less than $.5 \times 10^{-4}$. We have only retained four digits after the decimal point although the computation was carried out with 15 digits in the mantissa (i.e., 15 digits after the decimal point).

If the vector $\mathbf{b} = [6 \quad 1]^t$ in the equation $A\mathbf{x} = \mathbf{b}$, where A is the foregoing matrix, then a solution of the consistent system is $\mathbf{x} = A^+\mathbf{b} = [1.2308 \quad .6923 \quad .9231]^t$, taking the arbitrary vector $\mathbf{z} = \mathbf{0}$ in the general solution.. Out of infinite possible solutions, this solution has the minimum norm. Another solution $\mathbf{x} = [1 \quad 1 \quad 1]^t$ has a norm greater than the foregoing norm. If we take, in the equation $A\mathbf{x} = \mathbf{b}$, $\mathbf{b} = [6 \quad 2.9]^t$ and

$$A = \begin{bmatrix} 3 & 2 & 1 \\ 1.5 & 1 & .5 \end{bmatrix}$$

we get an inconsistent system of equations. The least-squares solution (whose norm is also minimum) of this inconsistent system is $\mathbf{x} = [1.2771 \ \ .8514 \ \ .4257]^t$. This solution will not satisfy the equation as the equation has no solution because of inconsistency. But the sum of the squares of the residuals, viz., $||A\mathbf{x} - \mathbf{b}||^2$ is a minimum as well as the norm of the vector $\mathbf{x}$, viz., $||\mathbf{x}||$ is also a minimum. *The minimum norm least squares solution* $\boldsymbol{x}$ *as well as the minimum norm least squares inverse* A^+ *are both unique. These are very useful in solving linear least-squares problems which arise in many physical problems including time-series analysis.*

We will discuss the error of a solution vector as well as that of an inverse computed by noniterative as well as other iterative algorithms later in a subsequent chapter.

2.2.12 Error in x of Ax = b in noniterative algorithms with nonsingular A

Consider the linear system $A\mathbf{x} = \mathbf{b}$, where A is nonsingular. It may be seen that *the nonsingularity of A mathematically implies that (i) the matrix A is square, (ii) it has linearly independent rows as well as linearly independent columns, (iii) the equation* $A\boldsymbol{x} = \boldsymbol{b}$ *is consistent,and (iv)* $A\boldsymbol{x} = \boldsymbol{b}$ *has a unique solution*. Let X be an approximate inverse of A and $\mathbf{z} = X\mathbf{b}$ be the approximate solution vector of the equation $A\mathbf{x} = \mathbf{b}$. Let the right-hand side residual be $Y = I - AX$ and the left-hand side residual be $Y = I - XA$. Choose that Y for which $||Y||$ is smaller. Let $\mathbf{r} = \mathbf{b} - A\mathbf{z}$ be the residual vector. If $||Y|| < 1$ then the *absolute error* in the approximate solution vector $\mathbf{z}$ can be given by the following inequality.

$$||\mathbf{r}||/||A|| \le ||A^{-1}\mathbf{b} - \mathbf{z}|| \le (||X|| \times ||Y|| \times ||\mathbf{b}||)/(1 - ||Y||).$$

The *relative error* in $\mathbf{z}$ is, when $||Y|| \le 1$, given as follows.

$$||A^{-1}\mathbf{b} - \mathbf{z}||/||A^{-1}\,\mathbf{b}|| \le ||Y||.$$

Both the foregoing errors could be used for any *noniterative* algorithm. For fixed-point iterative algorithms, both the relative as well as the absolute errors are obtained just by considering the most recent solution vector as well as the one just preceding it.

Consider the following *physical problem* (Figure 2.1). Find the current in each of the five resistors of the following electric circuit.

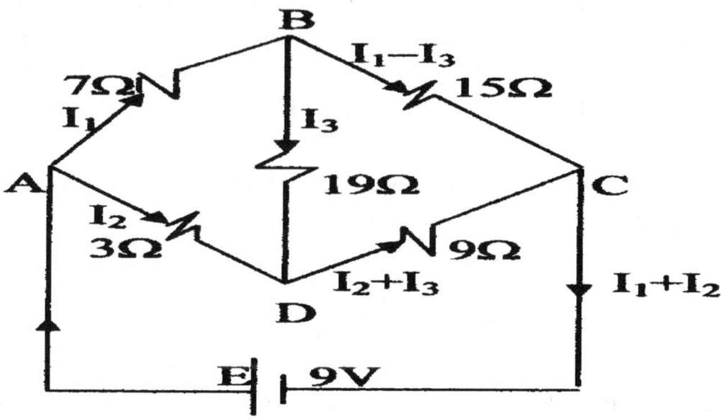

Figure 2.1 An electric circuit

Applying Kirchoff's Second Law (i.e., Voltage Law: In any closed circuit or mesh, the algebraic sum of products of currents and resistances (voltage drops) plus the algebraic sum of all the electromotive forces (emf's) in that circuit is zero) for each of the three meshes ABDA, BCDB, and ABCEA, we get

$$-7I_1-19I_3+3I_2=0, \text{ i.e., } -7I_1+3I_2-19I_3=0 \quad \text{(for Mesh ABDA)}$$
$$-15(I_1-I_3)+19I_3+9(I_2+I_3)=0, \text{ i.e., } -15I_1+9I_2+43I_3=0 \quad \text{(for Mesh BCDB)}$$
$$-7I_1-15(I_1-I_3)+9=0, \text{ i.e., } -22I_1+15I_3=-9 \quad \text{(for Mesh ABCEA)}$$

In matrix notation, these three equations can be written as $A\mathbf{x} = \mathbf{b}$:

$$\overset{A}{\begin{bmatrix} -7 & 3 & -19 \\ -15 & 9 & 43 \\ -22 & 0 & 15 \end{bmatrix}} \overset{\mathbf{x}}{\begin{bmatrix} I_1 \\ I_2 \\ I_3 \end{bmatrix}} = \overset{\mathbf{b}}{\begin{bmatrix} 0 \\ 0 \\ -9 \end{bmatrix}}$$

Hence $\mathbf{x} = X\mathbf{b} = [\, I_1 \;\; I_2 \;\; I_3]^t = [0.3930\text{A} \;\; 0.7677\text{A} \;\; -0.0236\text{A}]^t$, where

$$X = \begin{bmatrix} -0.0197 & 0.0066 & -0.0437 \\ 0.1049 & 0.0761 & -0.0853 \\ -0.0288 & 0.0096 & 0.0026 \end{bmatrix}$$

is the approximate (computed) inverse of A.

The relative error bounds (Fitzgerald 1970; Krishnamurthy and Sen 2001) in the computed solution vector **x** are given by $||A^{-1}\mathbf{b} - \mathbf{z}||/||A^{-1}\mathbf{b}|| \leq ||Y|| = ||I-AX|| = 1.3132 \times 10^{-16}$, where $\mathbf{z} = X\mathbf{b}$ is an approximate solution vector of $A\mathbf{x} = \mathbf{b}$, which implies that the quality of the result is excellent.

2.3 Error-free environment/quantity — How far is it possible

In the real world, problems dealing *with real quantities cannot be represented error-free in general unless, however, we deal with discrete quantities such as 50 balls, 20 mangoes, 10 beams, 1 thoudand people.* Since any measuring instrument that has associated with it a fixed order of nonzero error and that is used to measure a nondiscrete real quantity cannot produce error-free magnitude of a quantity. Thus *error is an integral part of any magnitude of a nondiscrete real quantity.* A one liter container that is used to measure milk from a big container of 25 liters has associated with it a fixed measuring error of, say, 5 gm. This means that the exact quantity of the milk that is measured by the small container will lie in [995 gm, 1005 gm]. This exact quantity is never known nor can it be known by any means. We can, however, have a higher precision measure where error or, equivalently error-bounds, can be made smaller, say 0.5 gm but this can never be made zero.

2.3.1 Even the discrete quantities may not be expressed error-free

We may/need not express even the discrete quantities exactly when the quantity is very large. Sometimes, exact discrete quantity which is certainly an integer cannot be easily obtained/counted. For example, Erythrocytes (red blood cells — RBC) of a healthy man is 4.71 million per cubic millimeter (mm^3), where the range of these cells for nondiseased people could be 3.9–5.8 million/mm^3 with mean 4.9 million/mm^3. The number of RBC is certainly not error-free. Depending on the measuring/counting device the error could be 0.5% or more or less. The number of Leukocytes (white blood cells — WBC) of a healthy man could be 7000 per mm^3 with range 3600-9900 per mm^3 and with mean 6800 per mm^3. Even this much smaller count is often not exact. The other haematological factors Neutrophils (42%-72% with mean 57%), Lymphocytes (20%-48% with mean 34%), Eosinophils (0%-20% with mean 9%), Basophils (0%), and Monocytes (0%-2% with mean 0.7%) of human blood are each measured in terms of %WBC. For example, 56, 36, 7, 0, 1, respectively for the foregoing five factors could be a reading for a human being. All these %WBC are not error-free, in general.

2.3.2 Error-free computation: Any error in the result?

Error-free computations (Gregory and Krishnamurthy 1984) can be carried out on rational numbers/integers (subject to the precision of the machine) as inputs when a deterministic algorithm that uses an error-free arithmetic, say multiple modulus residue arithmetic or a p-adic arithmetic or a floating-point modular arithmetic (Lakshmikantham et al. 1995, 1997, 2000; Sen and Howell 1992; Sen and Jayaram 1980; Sen and Shamim 1978a, 1978b; Sen 1980; Venkaiah and Sen 1987, 1988) involves a finite number of add, subtract, multiply, and divide operations. In all these arithmetic, intermediate number growth does not exist. The integer arithmetic, the rational arithmetic, or the rational complex arithmetic can also be used for error-free computation but these arithmetic are not free from intermediate number growth and hence are rarely used. The outputs of error-free computation are totally error-free ***if we assume that the input data are exact***. Otherwise, errors in input data will be often magnified in the output results. ***No error computation/analysis in the outputs when the error-bounds are specified for the input data has yet been reported in error-free computation.***

2.4 Error analysis

With the advent of digital computers in 1940s, error analysis (ea) became an important area of research. Gauss reduction method for solving $n \times n$ linear system $Ax = b$ became one of the algorithms to be explored. All the early error analyses were of forward type (discussed below). One of the important results was due to Hotelling who felt that errors in solving $Ax = b$ might build up 4^{n-1}. Subsequently, Bergmann, Montgomery, and von Neumann felt, in 1946, about the Gauss reduction method: An error injected at any stage affects all succeeding computations and may become greatly magnified. Such a feeling had a tint of pessimism. A sound assessment of the nature of the problem was not critically done. There were, however, fundamental limitations on the attainable accuracy because of the inherent errors in the floating-point/fixed-point arithmetic in a finite precision machine. Such errors will be discussed in a subsequent chapter.

Basic arithmetic operations on a digital computer are practically always subject to rounding or truncation errors. The cumulative effect of all these errors is of concern to us. There are two main methods of error analysis — ***forward error analysis (fea)*** and ***backward error analysis (bea)***. These methods are illustrated by considering the Gauss reduction method (Krishnamurthy and Sen 2001) for solving an $n \times n$ linear system $A\mathbf{x} = \mathbf{b}$. Assuming leading minors of all orders of the $n \times n$ matrix A nonzero, the original system is transformed (under elementary transformations) successively to equivalent[12] systems $A^{(r)}\mathbf{x} = \mathbf{b}^{(r)}$, r = 1(1)n − 1. The matrix $A^{(n-1)}$ is upper triangular. The solution is found by back-substitution.

In fea, the following strategy is used. Because of rounding errors, the computed system $A'^{(r)}\mathbf{x} = \mathbf{b}'^{(r)}$ differs from the error-free system $A^{(r)}\mathbf{x} = \mathbf{b}^{(r)}$. If the method is stable, then $||A'^{(r)} - A^{(r)}||$ and $||\mathbf{b}'^{(r)} - \mathbf{b}^{(r)}||$ will be small. With sufficient ingenuity, error bounds will be found.

In bea, the following strategy is used. If the algorithm is stable, the computed solution $\mathbf{x}'$ is the exact solution of some system $(A + E)\mathbf{x}' = \mathbf{b} + \mathbf{e}$, where E and **e** are relatively small. In fact, there will be an infinite number of E's and **e**'s, of which $\mathbf{x}'$ is the exact solution. A successful bea will obtain satisfactory bounds for the elements of E and **e**. The bea puts the errors produced during the computation in the same footing as those arising from the data. Consequently, when the initial data is in itself erroneous (inexact), no additional problem arises. For further information, the reader may refer (Wilkinson 1963, 1965; Forsythe and Moler 1967).

2.4.1 A priori versus a posteriori error analysis

The main function of the a priori error analysis is to show if an algorithm is stable. If it is not then the job of the analysis is to determine the reasons for its instability. In this analysis, we do not assume that a solution has already been obtained.

Having obtained an approximate solution, we can obtain sharper backward error bounds in a posteriori error analysis. Let λ and **x** be the computed eigenvalue and the corresponding normalized eigenvector ($||\mathbf{x}|| = 1$) of the n × n matrix A. If we write the residual $\mathbf{r} = A\mathbf{x} - \lambda\mathbf{x}$ as $(A - \mathbf{r}\mathbf{x}^H)\mathbf{x} = \lambda\mathbf{x}$ then the computed λ and computed **x** are exact for the matrix $A - \mathbf{r}\mathbf{x}^H$. The matrix A has an eigenvalue in the interval $[\lambda - ||\mathbf{r}||, \lambda + ||\mathbf{r}||]$ if A is Hermitian.

When solving linear system $A\mathbf{x} = \mathbf{b}$, we can compute the residual $\mathbf{r} = A\mathbf{x} - \mathbf{b}$, where **x** is the computed solution vector. An improved solution $\mathbf{x} - \Delta\mathbf{x}$ of the system can then be obtained by solving the system $A\Delta\mathbf{x} = \mathbf{r}$. Further improvement (iterative refinement) can be achieved by replacing **x** by $\mathbf{x} - \Delta\mathbf{x}$ and solving $A\Delta\mathbf{x} = \mathbf{r}$.

2.5 Limitation of interval arithmetic and significant digit arithmetic

We have already mentioned about the popular floating-point arithmetic in Section 2.4. Its error aspects will be discussed in a subsequent chapter. If we

---

[12] Equivalent in the sense that the solutions of the systems $A^{(r)}\ \mathbf{x} = \mathbf{b}^{®}$ r = 1(1)n are all equal.

are able to obtain error bounds for computed quantities at any stage of computation, it would indeed be very desirable. With this idea in view, *interval arithmetic* was designed. In this arithmetic an exact quantity q (usually not known) is represented as an ordered pair $[q_t, q_u]$ of floating-point numbers such that $q_t \le q \le q_u$. The computation that is carried out using the interval arithmetic (Moore 1966) is essentially the analog of what had been done algebraically in the Hotelling-type forward error analysis. The intervals grow very large. The results achieved are pessimistic. The computed quantities in stable algorithms need not be close to those arising in exact computation.

In *significant digit arithmetic*, computation is not done using the normalized floating-point numbers. This is because the zeros introduced are non-significant when cancellation takes place. The possibilities of significant digit arithmetic have been studied and well exploited by Metropolis and Ashenhurst.

It was realized that these arithmetic do not provide an automatic answer to error analysis/computation. Consequently, their use in science and engineering is restricted. However, these arithmetic still have an important place in theoretical error analysis.

2.6 Visualization of error

For a fixed-point iteration scheme or an iterative refinement scheme, we can draw the graph depicting how the error shrinks with the increasing number of iterations when the scheme converges. There are other two important possibilities, viz., the oscillation (sometimes oscillatory convergence) of iterates and divergence of iterates. In the case of divergence, the error depicted in the graph increases with the increase of the number of iterations. We will illustrate the visualization of error shrinking using the following example.

Example Visualize the error in computing a root of the equation $x^2 - 25 = 0$ using the Newton scheme

$$x_{i+1} = x_i - (f(x_i)/f'(x_i)),\ i = 0, 1, 2, \ldots, \text{ till } |x_{i+1} - x_i|/|x_{i+1}| \le 0.5 \times 10^{-4}.$$

where $f'(x) = df/dx$ and the initial approximation x_0 is chosen as 500 (too far away from the root 5).

Here the successive computed roots are

$$\mathbf{x} = [x_1\ x_2\ x_3\ x_4\ x_5\ x_6\ x_7\ x_8\ x_9\ x_{10}\ x_{11}]$$
$$= [250.0250\ \ 125.0625\ \ 62.6312\ \ 31.5152\ \ 16.1542\ \ 8.8509\ \ 5.8377\ \ 5.0601\ \ 5.0004\ \ 5.0000\ \ 5.0000]$$

for the iterations

$$iter = [iter_1 \ \ iter_2 \ldots iter_{11}] = [1 \ 2 \ 3 \ 4 \ 5 \ 6 \ 7 \ 8 \ 9 \ 10 \ 11].$$

The relative error vector is

$$er = [er_1 \ er_2 \ er_3 \ er_4 \ er_5 \ er_6 \ er_7 \ er_8 \ er_9 \ er_{10} \ er_{11}] = [0.9998 \quad 0.9992 \quad 0.9968 \quad 0.9873 \quad 0.9509 \quad 0.8251 \quad 0.5162 \quad 0.1537 \quad 0.0119 \quad 0.0001 \quad 0.0000],$$

where $er_1 = |x_1 - x_0|/|x_1|$, $er_2 = |x_2 - x_1|/|x_2|$, . ., $er_{11} = |x_{11} - x_{10}|/|x_{11}|$

The relative error graph (Figure 2.2) depicting iteration versus relative error is as follows.

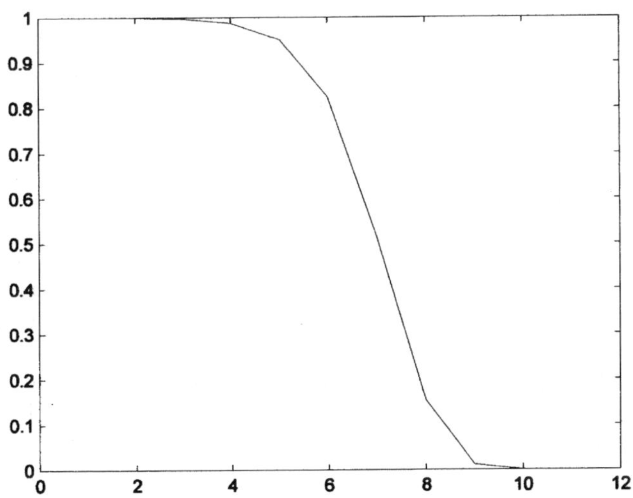

Figure 2.2: Iteration versus relative error in Newton scheme (order of convergence is 2) for solving the equation $x^2 - 25 = 0$ with initial approximation $x_0 = 500$; error reduces sharply after 5 iterations.

The relative error shrinks fast after five iterations. The computed root, viz., 5.0000 is correct at least up to 4 significant digits as required in the scheme.

We will present the error computation and visualization in subsequent chapters when we deal with specific algorithms.

2.7 Mathematical error versus computable error

Error-bounds (for an algorithm) derived mathematically, wherever possible, may not always be computable, although such mathematical error-bounds could provide better understanding of the error produced by the concerned algorithm. Consider, for example, the Largange interpolation formula (Krishnamurthy and Sen 2001). The problem is as follows. Given the table (x_i, y_i) $i = 0(1)n$, where x_i may or may not be equally spaced, get an n-th degree polynomial $y(x)$ that passes through all the $n+1$ points (x_i, y_i). This polynomial is an approximation of the function $f(x)$, which coincides with the polynomial at (x_i, y_i) $i = 0(1)n$. The mathematical form of a relative error in the formula (Krishnamurthy and Sen 2001) is given by

$$E_r(x) = [|x_n - x_0|^{n+1} \max |f^{n+1}(\xi)|/(n+1)!]/y(x), \quad x_0 \le \xi \le x_n$$

where $f^{n+1}(x)$ is the (n+1)-st derivative of the function $f(x)$ with respect to x, ξ is a value in $[x_0, x_n]$ so that this derivative will be maximum in magnitude. The value of ξ is not readily known; nor do we need to know ξ. All that we have to know is the largest value (in magnitude) of the (n+1)-st derivative of $f(x)$ in $[x_0, x_n]$. ***The function f(x) is not analytically known***. If, for example, $f(x) = 2\cos^2(x) + \sin(x^2) - 1$ then $f(x)$ is considered analytically known. *If f(x) is analytically known then there is usually no need to do interpolation.* One can directly evaluate f(x) for the given value of x. Hence the foregoing mathematical error formula is not of much use in practice, particularly in the age of extensively available personal computers (PCs). However, during the pre-computer days (before 1940s) or even during the early computer days (1940s – 1960s), we used to use mathematical tables (Abramowitz and Stegun 1965) for trigonometric (sine, cosine), special (Bessel, Legendre), probability (chi-square, F, Gaussian) functions. A mathematical table consists of function values at only discrete points. To know the function value at a point not in the table, interpolation used to be employed. Since the analytical function is known/specified for the table, the foregoing mathematical relative error formula can be used to compute the error-bounds. To-day the 'mathematical tables' part existing in the foregoing excellent book edited by M. Abramowitz and I.A. Stegun is practically obsolete because readily available built-in functions as well as extensively used software packages such as MATLAB, MATHEMATICA provide the required value of the function just by pressing a couple of keys on a PC.

A computable error in the case of Lagrange interpolation is one determined by an sma (sufficiently more accurate) value of the function along with (already computed) its la (less accurate) value. For example, if we want to know the error in the value $y = y_t$ computed by the Lagrange linear interpolation formula at a given point x then we may use the Largange quadratic interpolation to compute the value $y = y_q$ at the same point x. The value y_q could be considered the sma quantity and the value y_t the la quantity. The relative error in y_t is then $|y_q - y_t|/|y_q|$. We have assumed that the function defined by the table of values (x_i, y_i) $i = 0(1)n$ is analytic and well-conditioned, i.e, it is not at all sharply fluctuating.

The numerical interpolation is never done to produce a polynomial of degree n by considering all the n+1 points for large n. Usually, we consider linear or quadratic or cubic interpolation. We will be concerned with the computable error instead of mathematical error most of the time in algorithms to be discussed in subsequent chapters.

2.8 Confidence versus error

In numerical computation for a deterministic algorithm — iterative or noniterative — the relative error that is computed should/will be the error about which we are 100% confident. Consider, for example, the error (i.e., the relative error bounds) computed in the Newton scheme in Section 2.2.5 to find the square-root of the number 25. We are 100% confident that the exact solution (exact square-root of 25) has the relative error, say, $E_{r1} = 0.9992$, i.e., the exact square-root lies in the interval [0.2000, 499.6500] when we take $x_2 = 125.0625$ as the sma quantity and $x_1 = 250.0250$ as the la quantity. In this example, $E_{r6} = |x_7 - x_6|/|x_7| = |5.8377 - 8.8509|/|5.8377| = 0.5162$, where $x_7 = 5.8377$ is the sma quantity and $x_6 = 8.8509$ is the la quantity. We are 100% confident that the exact solution lies in $[x_6 - E_{r6}x_6, x_6 + E_{r6}x_6] = [4.2824, 13.4194]$. ***Since the confidence level is always 100% in the numerical computation concerning deterministic algorithms, we do not explicitly state the confidence level while specifying the error.***

In contrast, in numerical computation for a probabilistic/stochastic algorithm — such as a genetic algorithm or an algorithm based on evolutionary approach (Goldberg 2000, Koza 1998a, 1998b), an algorithm based on a probability distribution function (normal or t- or chi-square), a probabilistic OR (Operations Research) algorithm (Winston 1994), a PAC (probably approximately correct) concept learning algorithm (Vidyasagar 2003) — the error that we compute is always at a specified confidence level, say 95% or 99%. We do not specify 100% confidence level for error computation for such randomized (stochastic) algorithms. If we specify the confidence level as 100% here then the interval specified by the concerned error-bounds would be too large (and sometimes even $\propto$) to be meaningful/useful in the real-world situation. Most often, we specify the confidence level as 95%. Depending on the given context, this level could be taken as 90% or 99% and never 100%. However, the higher the confidence level is, the larger (non-sharper) the interval (larger error-bound – smaller error-bound) of the error is. We will consider the confidence-error relationship later in this book while dealing with randomized algorithms.

It can be observed that a problem for which we have a polynomial-time deterministic algorithm could be solved using a probabilistic algorithm. For example, a linear program *Minimize* $\mathbf{c}^t\mathbf{x}$ *subject to* $A\mathbf{x} = \mathbf{b}$, $\mathbf{x} \geq \mathbf{0}$ (null column vector) which could be solved using Karmarkar's projective transformation algorithm (Karmarkar 1984) in polynomial-time ($O(n^{3.5})$ operations) could also be solved using a randomized algorithm (Sen 2001).

It may be seen that probabilistic algorithms are polynomial-time (fast) while the corresponding deterministic algorithms (when these exist) could be polynomial-time or exponential (combinatorial)-time. The Monte Carlo method (Krishnamurthy and Sen 2001) to integrate an analytical function of a single variable or multiple variables with specified limits of integration is a polynomial-time randomized[13] algorithm. A deterministic polynomial-time method is the Simpson's 1/3 rule.

Yet another example of probabilistic algorithm is the *simulated annealing algorithm (SAA)* (Press et al. 1994) to solve the travelling salesman problem (TSP). The TSP is to find the shortest (costwise) cyclical itinerary for a travelling salesman (TS) who must visit N cities, each only once, with positions (x_i, y_i) $i = 1(1)N$ and return finally to his city of origin.

The deterministic algorithm to evaluate $(N - 1)!$ paths to obtain the exact (globally) shortest path is combinatorial which can be shown to be exponential as follows. From the Sterling formula, we have $(N - 1)! = (\sqrt{(2\pi)})e^{kN}$, where $N > 1$, $k = [(N - 0.5)\log_e(N - 1) + 1 - N + \theta/(12(N - 1))]/N$, $0 < \theta < 1$. (For large N, we have $k = (\lim_{N\to\infty}\log N) - 1$). Using MATLAB in all the computations here, for N = 20 and letting $\theta = 0.5$, we get

(N – 1)! = 1.216451004088320e+017,
k=((N–.5)*log(N–1)+1–N+.5/(12*(N–1)))/N=1.92093765381009 $(\sqrt{(2\pi)})e^{kN}$
= 1.213786762476202e+017.

The magnitude of the factorial function and that of the exponential function are comparable. Even to find the shortest path for the TS to travel only 20 cities, 1.216451004088320e+017 paths have to be evaluated! To travel 200 cities by the shortest path, 199! possible paths need to be evaluated by the deterministic algorithm!! Having done this enormous amount of evaluation, we are 100% confident that the shortest path produced is error-free. The SAA is a reasonable choice although the probability of the solution to have

[13] The Monte Carlo method is randomized because it uses random numbers and sometimes referred to as deterministic randomized algorithm since the more the uniformly distributed random hits are the better would be the integration value (usually).

no error is not 0, i.e., our confidence in the exactness of the solution is not 100%. Even a procedure for verification whether the solution is truly the shortest path or not is not polynomial-time.

The foregoing derterministic algorithm is slow and too expensive and hence is not used in solving real-world problems. The SAA developed by N.Metropolis in 1953 is a probabilistic algorithm which is polynomial-time and hence fast. While one may not be 100% confident that the output of the SAA is the shortest path, one would reasonably believe that the output is a path not very much different from (longer than) the true shortest path.

We will discuss the complexity of an algorithm, that tells us if the algorithm is fast (polynomial-time) or slow (exponential-time) and the concerned cost of computing resources (computing time/amount of computation using one or more processors and storage used) in Chapter 3.

2.9 Error-bound is non-decreasing while actual error need not be

It is interesting to note that an error-bound will usually go on increasing with the computation somewhat like entropy in thermodynamics. The more the amount of computation is, the larger the error-bound will be. In case of a computation, say multiplication of the result by an exact quantity (say, 1 or 1.5) might not increase the error-bound. But certainly it will not decrease the bound. However, a lengthy quantity, such as 2.32456298, could increase the error-bound for a fixed (word-length) precision machine, even if the quantity is exact.

The actual error, on the other hand, may decrease with the increase in the amount of computation. This is because the error could occur either on the negative side or on the positive side. Consequently the cumulative/resultant effect could nullify the error partially or fully.

Consider, for example, a rectangular water tank. Suppose that its exact length is $t = 3$ m (meter), exact height $h = 1$ m and exact width is $w = 2$ m. Then the exact volume of the tank is $v = t \times h \times w = 6\ m^3$. Let the device for measuring t, h, and w have an order of error 0.5 cm. Then the relative errors in measuring t, h, and w are 0.001667, 0.005, 0.0025. Consequently, the relative error bound in v is $0.001667+0.0050+0.0025 = 0.009167$ since, in multiplication, the relative errors are added (Krishnamurthy and Sen 2001). However, if the device measures the length t as 300.2 cm, the height h as 99.9 cm, and width w as 200.1 cm, then the volume v will be $6.000995\ m^3$ and the corresponding relative error is .0001658 which is much smaller than the foregoing relative error bound, viz., 0.009167. Note that a relative error bound should be such that the exact quantity must lie in the interval specified by the relative error.

2.10 Stability and error

Stability, specifically numerical stability, and error are closely and inversely related in the sense that if the stability is more in a domain then the error is less and vice versa. The term *stability* occurs extensively in mathematical science, more specifically in differential equations (both partial and ordinary). Significant amount of work (Lakshmikantham and Trigiante 2002 and the references mentioned in this monograph) has been done and is still being done in this area. To provide an *informal definition*, consider solving numerically a finite difference equation (FDE) corresponding to a partial differential equation (PDE) using a finite precision arithmetic, say, 15 digit floating-point arithmetic. Let α be the exact solution of the FDE, where the computation is assumed to have been done using the infinite precision (infinite word-length) machine. Each arithmetic operation (add/subtract/multiply, or divide operation) in the finite difference equation would involve some rounding error when the foregoing finite precision floating-point arithmetic is used. Hence the computed solution (produced by the machine) will not be α but α'. The finite difference scheme is ***stable*** if the cumulative effect of all the rounding errors is negligible in comparison with the solution α.

Mathematical definition of stability Let, at each (rectangular) mesh point N_{ij}, e_{ij} be the error introduced and $|e_{ij}| < \varepsilon$, where ε is a small positive number. A finite difference scheme is stable (Krishnamurthy and Sen 2001) if

(i) $\max_{i,j}\left|\alpha_{ij} - \alpha'_{ij}\right| \to 0$ as $\varepsilon \to 0$

and

(ii) $\max_{i,j}\left|\alpha_{ij} - \alpha'_{ij}\right|$ does not increase exponentially with i and j.

Condition (ii) is necessary because the errors e_{ij} may not decrease exponentially with i, j but may continue to be a linear combination of the initial errors. In such a case, the scheme is accepted as stable if the cumulative error (sum of all errors) is much smaller than the solution α.

While it is not possible to obtain the exact value of $|\alpha_{ij} - \alpha'_{ij}|$ at each mesh point N_{ij}, an estimate of $|\alpha_{ij} - \alpha'_{ij}|$ can be obtained in a few special cases. The computed solution is always more accurate than what the estimate shows since the stability analysis considers the error bounds while obtaining an estimate.

Observe that the stability is not directly associated with the solution of a PDE. Also, note that the total error in solving a PDE is given as $(\alpha'' - \alpha') = (\alpha'' - \alpha) + (\alpha - \alpha')$, where α'' = the exact solution of the PDE, α = the exact solution of the FDE, α' = the computed solution (with rounding errors) of the FDE, $\alpha'' - \alpha$ = the truncation (discretization) error, and $\alpha - \alpha'$ = the stability error. One can see that the discretization error is usually dominant in a stable and convergent scheme.

One of the two methods, viz., the matrix method (Smith 1965) and the finite Fourier series method (O'Brien et al. 1951; Krishnamurthy and Sen 2001) could be used to analyze the stability of an implicit or an explicit finite difference scheme and to investigate the growth of errors in the computations needed to solve an FDE. The finite Fourier series method is simpler than the matrix method since it does not need the boundary conditions. However, these methods are not easily applicable to any arbitrary system of FDEs. Numerical error (bounds) at a mesh point, on the other hand, can be computed by computing sma value of the mesh point and the la value of the mesh point. These computations could require twice the time on a computer. But it is necessary if we desire to know the quality of the solution.

Different kinds of stability Let B_δ denote the open ball with its radius δ and its centre at y = 0. Consider the (implicit or explicit) FDE.

$$y_{n+1} = f(n, y_n),\ y_{n0} = y_0.$$

The solution y = 0 of the foregoing FDE will be called (Lakshmikantham and Trigiante 2002)

a. *stable* if, there is a $\delta(\varepsilon, n0) > 0$ so that for any $y_0 \in B_\delta$, the solution $y_n \in B_\varepsilon$ for a given $\varepsilon > 0$,
b. *uniformly stable* if the solution is stable and δ can be selected independent of n0,
c. *asymptotically stable* if it is stable and attractive[14],
d. *uniformly asymptotically stable* if it is uniformly stable and uniformly attractive[15],
e. *globally asymptotically stable* if it is asymptotically stable for all starting points y_0.
f. *uniformly exponentially stable* if there are a positive δ, a positive α, and an $0 < \eta < 1$ so that $\|y_0\| \le \alpha\|y_0\|\eta^{n-n0}$ whenever $y_0 \in B_\delta$.

Further, the solution could also be defined l_p-*stable*[16] as well as *uniformly* l_p-*stable* besides *totally stable* (Lakshmikantham and Trigiante 2002). The solution y = 0 of the FDE $y_{n+1} = f(n, y_n)$ will be totally stable if for every $\varepsilon > 0$, there are two numbers $\delta_1(\varepsilon) > 0$ and $\delta_2(\varepsilon) > 0$ so that every solution $y(n, n_0, y_0)$ of the FDE $y_{n+1} = f(n, y_n) + R(n,y_n)$, where R is a bounded Lipschitz function in B_a and R(n, 0) = 0.

---

[14] *attractive* if there exists a $\delta(n0) > 0$ so that $\lim y_n = 0$ as $n \to \infty$ for $\in B_\delta$,

[15] *uniformly attractive* if it is attractive and δ can be selected independent of n0,

[16] If a solution y = 0 is exponentially stable then it is also l_p-stable.

The foregoing definition of different forms of stability provides us a better insight of the problems of solving FDEs and possibly would help us in correlating error with stability. For further details on stability of various numerical problems, refer Butcher (1975), Burrage and Butcher (1979), Cash (1979), Dahlquist (1963, 1975a, 1975b, 1978,1983), Dahlquist et al. (1083), Elman (1986), Hurt (1967), LaSalle (1979), Lena and Trigiante (1982, 1990), Mattheij (1984), Melvin (1974), Ortega (1973), and Sugiyama (1971).

Bibliography

Abramowitz, M.; Stegun, I. A. (eds.) (1965): *Handbook of Mathematical Functions with Formulas, Graphs, and Mathematical Tables*, Dover Publications, Inc., New York.

Burrage, K.; Butcher, J.C (1979): Stability criteria a implicit Runge-Kutta Methods, *SIAM J. Numer. Anal.* **15**, 46-57.

Butcher, J. C. (1975): A stability property for implicit Runge-Kutta methods, *BIT*, **15**, 358-61.

Cash, J.R. (1979): *Stable Recursions*, Academic Press, London.

Dahlquist, G. (1963): A special stability problem for linear multistep methods, *BIT*, **3**, 27-43.

Dahlquist, G. (1975a): Error analysis for a class a methods for stiff nonlinear initial value problems, *Num. Anal. Dundee* Spring Lect. Notes in Math., **506**, 60-74.

Dahlquist, G. (1975b): On stability and error analysis for stiff nonlinear problems, Part 1, *Report Trita-NA*-7508.

Dahlquist, G. (1978): G-stability is equivalent to A-stability, *BIT*, **18**, 384-401.

Dahlquist, G. (1983): Some comments on stability and error analysis for stiff nonlinear differential systems, preprint *NADA* Stockholm.

Dahlquist, G.; L. W. and O. Nevanlinna (1983): Stability of two-step methods for variable integration steps, *SIAM J. Numer. Anal.*, **20**, 1071-85.

Elman, H. (1986): A stability analysis of incomplete LU factorisation, *Math. Comp.*47, 191-217.

Fitzgerald, B.K. E. (1970): error estimates for the solution of linear algebraic system, *J. Res. Nat. Bur. Sts.,* **74B**, 251-310.

Forsythe, G.E.; Moler, C.B. (1967): *Computer Solution of Linear Algebraic Systems*, Prentice-Hall, Englewood Cliffs, New Jersey.

Goldberg, D.E. (2000): Genetic *Algorithms in Search, Optimization, and Machine Learning*, Addison-Wesley, Reading, Massachusetts.

Gregory, R.T.; Krishnamurthy, E.V. (1904): *Methods and Applications of Error-free Computation*, Springer-Verlag, New York.

Hurt, J. (1967): Some stability theorems for ordinary difference equations, *SIAM J. Numer. Anal.*, 4, 582-96.

Karmarkar, N. (1984): A new polynomial-time algorithm in linear programming, *Combinatorics,* 4, 373-395.

Koza, J.R. (1998a): G*enetic Programming: On the Programming of Computers by Means of Natural Selection*, MIT Press, Cambridge, Massachusetts.

Koza, J.R. (1998b): *Genetic Programming II: Automatic Discovery of Reusable Programs*, MIT Press, Cambridge, Massachusetts

Krishnamurthy, E.V. (1971): Complementary two-way algorithms for negative radix conversions, *IEEE Trans. Computers,*C-20, 543-550.

Krishnamurthy, E.V.; Sen, S.K. (2001): *Numerical Algorithms: Computations in Science and Engineering*, Affiliated East West Press, New Delhi.

Lakshmikantham, V.; Sen, S.K.; Sivasundaram, S. (1995): Computing polynomial root-clusters exactly and parallely, *Engineering Simulation (Amsterdam B.V. Published under licence by Gordon and Breach Science Publishers SA)*, 12, 291 – 313.

Lakshmikantham, V.; Sen, S.K.; Howell, G. (1996): Vectors versus matrices: p-inversion, cryptographic application, and vector implementation, *Neural, Parallel and Scientific Computations,* 4, 129-140.

Lakshmikantham, V.; Sen, S.K.; Maulloo, A. K.; Sivasundaram, S. (1997): Solving linear programming problems exactly *Applied Mathematics and Computation (Elsevier Science Pub. Co., New York)*, **81**, 69-87.

Lakshmikantham, V.; Sen, S.K.; Jain, M.K.; Ramful, A. (2000): $O(n^3)$ noniterative heuristic algorithm for linear programs with error-free implementation, *Applied Mathematics and Computation (Elsevier Science Inc., New York)*, **110**, 2000, 53-81.

Lakshmikantham, V.; Trigiante, D. (2002): *Theory of Difference Equations: Numerical Methods and Applications,* 2nd ed., Marcel Dekker, New York.

LaSalle, J.P. (1979): The stability of dynamical systems, *Regional Conference Series in Applied Mathematics*, SIAM.

Lena G.D.; Trigiante, D. (1982): On the stability and convergence of lines method, *Rend. Di Mat.*, **3**, 113-26.

Lena G.D.; Trigiante, D. (1990): Stability and spectral properties of incomplete factorisation, *Japan J. Appl. Math.*,**7**, 145-53.

Mattheij, R.M. (1984): Stability of block LU-decompositions of the matrices arising from BVP, *SIAM J. Alg. Dis. Math.*, **5**, 314-331.

Melvin, W. (1974): Stability properties of functional differential equations, *J. Math. Anal. Appl.*, **48**, 749-63.

Moore, R.E. (1966): *Interval Analysis*, Prentice-Hall, Englewood Cliffs, New Jersey.

O'Brien, G.G.; Hyman, M.A.; Kaplan, S. (1951): A stydy of the numerical solution of partial differential equations, *J. Math. Phy.*, **29**, 223-51.

Ogden, T. (2000): *Ghosts and Hauntings* (Chap 3), Alpha Books.

Ortega, J.M. (1973): Stability of difference equations and convergence of iterative processes, *SIAM J. Numer. Anal.*,10, 268-82.

Press, W.H.; Teukolsky, S.A.; Vetterling, W.T.; Flannery, B.P. (1984): *Numerical Recipes in C/Fortran*, Prentice-Hall of India, New Delhi.

Sen, S.K. (1980): Nonnegative integral solution of linear equations, *Proc. Ind. Acad. Sci. (Mathematical Sciences)*, **89,** 1, 25 – 33.

Sen, S.K. (2001): Linear program solver: evolutionary approach, *Proc. 46th Congress of ISTAM (International Meet)*, 75-84.

Sen, S.K. (2002): Error and computational complexity in Engineering, in *Computational Mathematics, Modelling, and Algorithms,* ed. J.C. Misra, Narosa Publishing House, New Delhi.

Sen, S.K.; Jayaram, N.R. (1980): Exact computation of a matrix symmetrizer using p-adic arithmetic, *J. Ind. Inst. Sci.,* 62A, 1980, 117 – 128.

Sen, S.K.; Prabhu, S.S. (1976): Optimal iterative schemes for computing Moore-Penrose matrix inverse, *Int. J. Systems Sci.*, 8, 748-753.

Sen, S.K.; Howell, G.W. (1992): Direct fail-proof triangularization algorithms for AX + XB = C with error-free and parallel implementations, **J.** *Appl. Maths. and Computation* (Elsevier Science Pub. Co., New York), **50,** 1992, 255 – 278.

Sen, S.K.; Shamim, A. A. (1978): An integer arithmetic method to compute generalized matrix inverse and solve linear equations exactly, *Proc. Ind. Acad. Sci.*, **87A,** 9, 161 – 168.

Sen, S.K.; Shamim, A.A. (1978): Integral solution of linear equations using integer arithmetic, *J. Ind. Inst. Sci.,* **60,** 3, 1978, 111 – 118.

Smith, G.D. (1965): *Numerical Solution of Partial Differential Equations*, Oxford University Press, Oxford.

Sugiyama, S. (1971): Difference inequalities and their applications to stability problems, *Lecture Notes in Math., Springer*, **243**, 1-15.

Swami Prabhavananda; Christopher, Isherwood (2002): *Patanjali Yoga Sutras,* Sri Ramakrishna Math, Chennai (The authors translated from Sanskrit — an ancient Indian language — with new commentary).

Swami Satyananda Saraswati (2000): *Four Chapters on Freedom: Commentary on the Yoga Sutras of Patanjali*, Yoga Publications Trust, Munger, Bihar.

Venkaiah, V. Ch.; Sen, S.K. (1987): A floating-point-like modular arithmetic for polynomials with application to rational matrix processors, *Advances in Modelling and Simulation*, **9,** 1, 1 – 12.

Venkaiah, V. Ch.; Sen, S.K. (1988): Computing a matrix symmetrizer exactly using modified multiple modulus residue arithmetic, *J. Computational and Applied Mathematics (Elsevier Science Publishers B.V., North-Holland)*, **21,** 27 – 40.

Vidyasagar, M. (2003): *Learning and Generalization: With Applications to Neural Networks*, 2nd Edition, Springer, London.

Wilkinson, J.H. (1963): *Rounding Errors in Algebraic Processes*, Her Majesty's Stationery Office, London.

Wilkinson, J.H. (1965): *The Algebraic Eigenvalue Problem,* Clarendon Press, Oxford.

Winston, W.L. (1994): *Operations Research: Applications and Algorithms,* Duxbury Press, Belmont, California.

Chapter 3

Complexity: What, Why, and How

3.1 Introduction

The word "complexity" is the noun of the word "complex" which means, according to Collins Gem English Dictionary, *made up of parts, complicated.* As a noun, complexity means *whole made up of parts, group of unconscious feelings that influences behaviour*. The word *complex* as opposed to the word *simple* implies difficult or complicated. So far as a human being is concerned, it might mean difficult to understand/grasp. So far as a computer is concerned, it would not certainly mean difficult to understand/grasp as the nonliving machine has no such feeling of easy/simple or difficult/complicated. For a common man it is difficult to grasp Maxwell's electromagnetic laws or quantum physics although the physical size (for representation on a piece of paper) of these laws or of the quantum physics is not very large. The grasping needs background knowledge and training as well as sufficient intelligence. This one kind of complexity related only to man.

There is yet another (second) kind of complexity which we will be concerned with. For a living computer such as a human being, the larger the number of things to be comprehended, the more difficult it is. For example, a common man can remember/grasp 7 ± 2 names by hearing them once. But, if 20000 names are read out to him, he would not be able to register almost all of them in his mind/brain. The problem of remembering 20000 names is a difficult/complex issue to a normal man while that of remembering 7 ± 2 names is a less complex (simple) issue to him. Such a problem is assumed nonexistent in a machine. The machine would be able to remember all the 20000 names by getting them only once. Further, with the passage of time, unlike a common man, it would not at all forget the names.

Keeping in mind this ***difference*** existing between the man and the machine and yet the ***analogy of the second kind***, we define the complexity, rather the computational complexity in the realm of digital computers, precisely as the amount of computation carried out by an algorithm before producing the required output for a specified input. So the complexity is expressed in terms of the function (polynomial or exponential) of the input size. The *amount of computation* is measured in terms of number of operations (arithmetic such as add, subtract, multiply, divide and nonarithmetic such as test, jump) involved in the execution of the program (algorithm). The more the amount of computation is, the larger is the computational complexity.

Different operations/instructions need different times to be executed. For example a multiply operation takes at least 1.2 times the time needed by an add operation. Even the same two operations, say two multiplications — one of the contents of two operands (existing in executable memory physically) at certain distances from the central processing unit (CPU) while the other of the contents of another two operands at certain different distances — will take different times for their execution. This is because the *fetch* microinstructions would take different times since the data (contents of operands) movement between the CPU and the memory would not exceed the speed of light; Light takes different times to travel different distances. Therefore, often we consider average time needed for the execution of an instruction. Here the execution of an instruction, say multiplication, consists of four microinstructions, viz., *fetch, decode, execute, and write back* (Donovan 1972).

Under these circumstances, the computational complexity could also be expressed as time complexity which is defined as the amount of time needed by the algorithm to produce the required output for a specified input.

3.2 Algorithm as Turing machine and algorithmic complexity

3.2.1 Godel's incompleteness theorem

D. Hilbert, a great German mathematician, proposed at the beginning of twentieth century, 23 problems which, he believed, needed to be solved in all parts (of *Hilbert's program*) to put solid logical foundation under all of mathematics (Whitehead and Russell 1910, 1912, and 1913). One of these problems, the *decision problem*, called for a step-by-step procedure — an *algorithm* — for deducing all the propositions that are true within any mathematical system (Glenn 1996).

The late nineteenth and early twentieth centuries, mathematicians under the inspiration of D. Hilbert had hoped to find a mechanical method for expressing and verifying all the mathematical truths arising from a set of axioms. One of the big mathematical goals was to reduce all of number theory to a formal axiomatic system. Like Euclid's geometry, such a system would start off with a few simple axioms that are almost indisputable, and would provide a mechanical way of deriving theorems from these axioms. Their hope was dashed when K. Godel, a brilliant Austrian logician, demonstrated in a proof that any part of mathematics at least as complex as arithmetic can never be complete (Godel 1931). No algorithm, howsoever large, can lead to sorting out all the true or untrue statements/information/equations within a system. He demonstrated that statements exist that cannot be derived by the rules of arithmetic proof. He, through his *incompleteness theorem*, showed that no method of proof could be subjected to mechanical verification as well as be powerful enough to prove all the theorems of elementary arithmetic.

Godel proved that, for any formal axiomatic system, there is always a statement about natural numbers which is true, but which cannot be proved in the system. Mathematics thus will never be the rigorous unshakable system which mathematicians dreamt of for ages. In other words, mathematics will always have some fuzziness near the boundary.

Consider, for example, Table 3.1 the Typographical Number Theory (TNT) which uses the following symbols, variables, numbers, axioms, and proof methods (Felder 1996).

Table 3.1 Symbols, variables, axioms, and proof methods used in TNT

Symbols	
(mathematical)	+, *, =
(logical)	~(not), v (or, E (there exists), A (for all)
(numbers)	0 (zero), S (successor of)
Variables	
Letter a followed by primes	(a, a′, a″, a‴, …)
Axioms (axiom strings)	
1:	Aa: ~Sa=0 (no negative number)
2:	Aa: (a+0)=a
3:	Aa:a′: (a+Sa′)=S(a+a′)
4:	Aa (a*0)=0
5:	Aa: Aa′☹a*Sa′)-((a*a′)+a)
Proof methods	
Rules (string manipulation rules	
1:	The string ~ ~ can be deleted
2:	For any variable a, the strings Aa: ~ and ~Ea: are interchangeable

Any string produced following axiom(s) and rules (manipulation) is a *theorem*. In other words, we have (Figure 3.1)

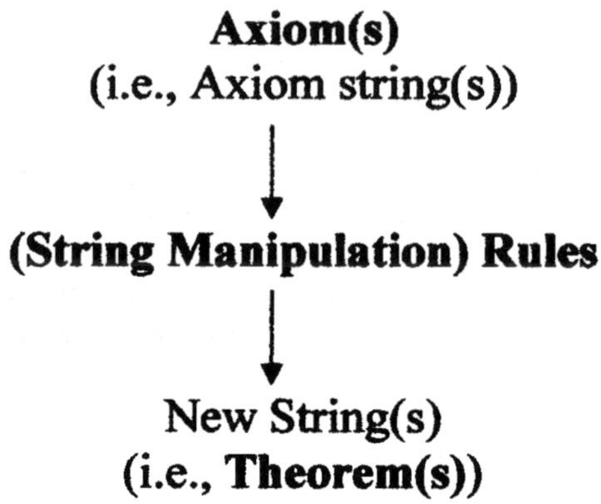

Figure 3.1: Generation of theorem(s) from rules with axiom(s) as Input(s)

Example Aa : ~Sa = 0 (Axiom 1) ~Sa : Sa = 0 (Rule 2)

Theorem: **S0 +S0 = SS0** ***Theorem***: **Aa : Aa′ : (a + a′) = (a′ + a)**

A TNT string ~Ea : a*a*a = SSS0 means that there does not exist any number a, such that a times a times a is 3, i.e., there is no cube-root of 3. This string is a *true* statement since TNT deals only with natural numbers. However, if we replace SSS0 by S0 in the foregoing string then the resulting string is a *false* statement.

It may be seen that any statement that one can make about natural numbers can be written in a TNT string. If such a statement is true then its TNT string can be obtained as a theorem from the axioms. If the statement is false then its converse can be derived from the axioms.

Consider the following example.

Sentence U: a = SSS0 * a – SSSSSS0
Sentence W: Sentence U is 3 times sentence U – 6.

Sentence U is neither true nor false as a is not specified. But the sentence W, called the arithmoquine (Felder 1996) of sentence U, is a false statement about a specific natural number. Here

U: The arithmoquine of a is not a valid TNT theorem-number.

W: The arithmoquine of sentence U is not a valid TNT theorem-number.

If we write sentence W as one big sentence without sentence U, we get the sentence Y:

Y: The arithmoquine of "The arithmoquine of a is not a valid TNT theorem-number" is not a valid TNT theorem-number.

Sentence Y is not a theorem of TNT: If the sentence Y is false then Y is not a theorem of TNT ⇒ there is a valid theorem (in TNT) which is false. If Y is true then Y is not a theorem of TNT ⇒ Sentence Y is true but it is not provable (in TNT). This is Godel's incompleteness theorem. Does this theorem imply the existence of facts that must be true but we are incapable of proving them?

3.2.2 Parallel between algorithm and theorem

It may be interesting to observe that there is a parallel between an algorithm (i.e., a formalized set of rules which can be mechanized) and a theorem in mathematics (Figure 2). The output in the case of an algorithm proves the validity of the algorithm while a proof does in the case of a theorem.

Input → Algorithm → Output (Result)

Input → Theorem → Output (Proof)

Figure 3.2: Algorithm versus theorem

3.2.3 Algorithmic undecidability

Can we devise an algorithm for carrying out any task? The answer is no. There are problems which are algorithmically undecidable. (Davis 1958). This algorithmic undecidability relates directly to the question whether there are statements in an axiom system that are not provable. There are indeed such statements in an axiom system that are neither proved nor disproved (Godel 1961; Nagel and Newman 1964). Similarly, there are tasks which are algorithmically undecidable, i.e., no algorithmic solution can be sought for these tasks.

3.2.4 Algorithm as a Turing machine and vice versa

The incompleteness theorem prompted the logicians to ask *What is an algorithm?* Several formal definitions were provided by Kleene, Church, Post, and Turing (Brady 1978; Clark and Cowell 1976; Manna 1974). All these definitions are equivalent and can be written as one definition: *Any algorithm can be expressed as a Turing machine*[1] *and any Turing machine expresses an algorithm.*

Turing developed his theoretical computational model in 1936. He based his model on how he perceived mathematicians think. Turing machine proved itself as the right hypothetical model for computation as the digital computers were designed and developed through 1940's and 1950's. The development of general purpose digital computers made possible the implementation and execution of complicated algorithms. Consequently the theory of computation became an area of great interest.

3.2.5 TM's incapability to account for complexity

However, the computability theory as developed by Turing and other logicians was not concerned with resource use and practicability. The basic Turing machine fails to account for the amount of time and memory needed by a computer — a critical issue even in those early years of computing. This issue to measure time and space as a function of the length of the input appeared in the early 1960's by Hartmanis and Stearns (Hartmanis 1994; Stearns 1994). Thus computational complexity came into existence.

Consider, for example, the problem of finding a subgraph[2] which is isomorphic[3] to a given graph (Harary 1972). All the known algorithms for this

---

[1] A Turing machine is a theoretical device with an infinite supply of paper tape marked-off as square regions. Such a machine is capable of performing just four actions, viz., *moving* the tape one square right or left, *placing* a mark on a square, *erasing* a mark and *halting*. Turing discovered that even a machine so simple as this can solve any problem for which an algorithm can be devised.)

[2] A graph is a collection of points, any pair of which may or may not be joined by a line. A subgraph of a graph G is a graph whose points and edges are all in G.

[3] Two graphs are isomorphic if there exists a 1-1 correspondence between their point sets which preserves adjacency. A graph or a directed graph (digraph) is represented by its adjacency matrix or adjacency structure. The $n \times n$ adjacency matrix for a graph on n vertices (points) is $A = [a_{ij}]$, where $a_{ij} = 1$ if the vertex v_i is adjacent to vertex v_j, i.e., $\{V_i, v_j)$ is an edge of the graph

problem have an execution time which increases exponentially with the increase in number of vertices in the graph. The execution time of any such algorithm is an exponential function (non-polynomial, i.e., polynomial of degree ∞) of the input size, viz., the number of vertices. All these algorithms are exponential-time and thus *slow* or inefficient. Nobody has so far discovered a polynomial-time, i.e., *fast* algorithm. Hence, although a problem can be solved in principle in the Turing sense, no efficient (fast) algorithm might exist. We now ask: What is algorithmic efficiency and how to measure algorithmic complexity?

3.2.6 Complexity as a function of input size

Although there are two kinds of complexity measures (Krishnamurthy and Sen 2001), viz., static[4] and dynamic[5], the much more important measures are the dynamic complexity measures. The dynamic measures give information about the resource requirement of the algorithm as a function of the size of the input data which need to be specified.

3.2.7 Worst and average case complexity measures

One way is to assume that the input data for a given problem is the worst possible. Such a worst case measure provides us an upper bound for (dynamic) complexity for real world use. Another approach is to assume that the input data is average. Such an average case complexity or, simply, average complexity measure provides us an average performance of the algorithm. One could also define the best case time complexity or, simply, the best time complexity which gives the lower bound for complexity.

Consider, for example, the quick sort algorithm to sort an array of n elements in some order (ascending or descending). The worst case time complexity of the algorithm is $O(n^2)$. This would happen if the pivot is always the greatest (the least) element at each recursive call at which the array is split into parts. The average time complexity of the quick sort is $O(n \times \log_2 n)$ which is also the best time complexity. In the average case the pivot has an equal probability of being the extreme (the greatest or the least) and not being the extreme. If there are p processors, where $p < n$, then the overall (average) computational complexity for a parallel quick sort is $O((n/p) \times \log_2 n)$.

---

and = 0 otherwise. The adjacency structure is the listing for each vertex of all other vertices adjacent to it.

[4] independent of the size and characteristics of the input data, e.g., program length gives the static complexity

[5] dependent on the input data, e.g., storage space and running time

The algorithms whose execution time is an exponential function of the size of the input, i.e., whose execution time grows exponentially as a function of the input size are not useful, in general. On the other hand, the algorithms whose execution time is a polynomial of the size of the input are considered efficient or fast for general application.

For example, the conventional matrix multiplication algorithm to compute the matrix C = AB, where $A = [a_{ij}]$, $B = [b_{ij}]$, and $C = [c_{ij}] = \Sigma a_{ik}b_{kj}$ are three $n \times n$ real matrices and the summation Σ is over k = 1 to n, is a polynomial-time algorithm needing n^3 multiplications and $n^2 (n - 1)$ additions besides the nonarithmetic operations such as branch, loop. This algorithm can also be termed as the $O(n^3)$ algorithm. We are more interested in knowing the highest order/degree of operation expressed in terms of n and not on lower order of operations. Also, we are concerned with truly large input size n and not with small n. Thus, n, n^2 are considered negligible with respect to n^3.

On the other hand, to solve switching (Boolean) equations, for the full adder that computes x + y + c,

$$\text{sum} = x'y'c + x'yc' + xy'c' + xyc$$
$$\text{carry} = x'yc + xy'c + xyc' + xyc$$

where sum = 1 and carry = 1 using the truth-table method, we need to evaluate the right-hand sides of the equations for $2^3 = 8$ ordered triplets of data for the 3 switching variables x, y, and c. These 8 triplets are (0, 0, 0), (0, 0, 1), (0, 1, 0), (0, 1, 1), (1, 0, 0), (1, 0, 1), (1, 1, 0), (1, 1, 1). The solution thus found is x = 1, y = 1, and c = 1. For n switching variables the method would need the evaluation of 2^n switching expressions. Hence the computational complexity of such a method is $O(2^n) = O(e^{n\log 2})$. Consequently, this method is exponential and so inefficient and is not useful in practice (for large n). A polynomial method for this (satisfiability) problem is yet to be discovered.

It may be observed that if the computational complexity of an algorithm is $O(\alpha n^3)$ to solve a particular type of problems and that of another algorithm for the same type is $O(\beta n^{2.9})$, where n is the input parameter, then the later algorithm is definitely more desirable for practical use, where we have assumed n sufficiently large. If $\alpha = 2$ and $\beta = 500$ then the first algorithm will perform better than the second (later) one for all n < 9.536743164062500e+023. In the real world situation, where $n < 9.5 \times 10^{23}$ (seems reasonably large), we should use the $O(2n^3)$ algorithm and not the $O(500n^{2.9})$ algorithm although the later one could be a sort of break through and of immense academic interest.

We have already seen that the algorithms that can be considered fast or efficient enough for general application are those whose execution time is a polynomial of the input size. Even among these algorithms, some are faster than others. A faster matrix multiplication is due to V. Strassen (Aho et al. 1974). For the input 2×2 matrices A and B, where n = 2, the conventional multiplication needs $n^3 = 8$ multiplications, $n^2(n-1) = 4$ additions (without distinguishing between addition and subtraction), and other control operations that exist in the program (for the conventional matrix multiplication). As seen earlier, the computational complexity is written as $O((n^2)^{3/2}) = O(n^3)$, since n^3 is the most dominant term. The conventional matrix multiplication is *fast* or *efficient* or, equivalently, *polynomial-time*. In Strassen method (Sen 2003), the product matrix C = AB needs $O(n^{\log 7}) = O(n^{2.8074})$ multiplications, where the base of the logarithm is 2. For n = 2, the method needs $2^{\log 7} = 7$ multiplications of numbers. For large $n = 2^k$, where k is a positive integer, one would be able to appreciate the gain in Strassen's method. It can be seen that multiplications are more important than additions (counting also subtractions and not distinguishing between them). Thus, even among these polynomial-time algorithms, some are faster than others. But for the purposes of our discussion, it is sufficient to distinguish the polynomial-time algorithms or, simply, ***polynomial algorithm*** as a class distinct from the exponential-time algorithms or, simply, ***exponential algorithms***. Such a classification makes the speed of an algorithm a property inherent in the algorithm itself independent of the property of the machine (computer). When the input size is sufficiently large, i.e., when the problem is sufficiently large, a polynomial algorithm executed on the slowest machine will find an answer (result) sooner than the exponential algorithm on the fastest machine.

Thus we see that the TM divides problems as solvable and unsolvable while the algorithmic complexity classifies the solvable problems into (i) those which can be solved using polynomial algorithms, (ii) those which cannot be solved by polynomial algorithms, i.e., which needs only exponential algorithms, and (iii) those for which no polynomial algorithms are known and for which the best available algorithms are exponential and yet for which no one has proved so far that no polynomial algorithms exist.

3.2.8 Examples on complexity measure

Linear system For solving a linear system $A\mathbf{x} = \mathbf{b}$, polynomial algorithms, such as the Gauss reduction method with partial pivoting (Krishnamurthy and Sen 2001) needing $O(n^3/3)$ operations, exist.

Travelling Salesman Problem For solving the travelling salesman problem (TSP) for N cities (Press et al. 1984), the only deterministic algorithm available is combinatorial, i.e., exponential (see Chapter 2). It requires $(N-1)! \approx (\sqrt{(2\pi)})e^{kN}$, where N is a sufficiently large positive integer and $k = (\lim_{N\to\infty} \log N) - 1)$.

Linear program: simplex and polynomial algorithm For solving a linear program (LP) *Maximize* $\mathbf{c}^t\mathbf{x}$ *subject to* $A\mathbf{x} = \mathbf{b}$, $\mathbf{x} \geq \mathbf{0}$, we did not have a polynomial algorithm till 1984 nor did one prove that a polynomial algorithm existed for an LP. It was N. Karmarkar (Karmarkar 1984) who first[6] showed through his projective transformation (interior-point) algorithm the existence of a polynomial algorithm needing $O(n^{3.5})$ operations for LPs. The only popular algorithm till then was the (exterior-point) simplex algorithm (SA) which was not polynomial although it behaved like one for most real world problems for decades (Dantzig 1963; Gass 1969, 1975; Luenberger 1973; Krishnamurthy and Sen 2001; Vajda 1975; Winston 1994). In a tableaux of the SA, a variable can enter into the basis (Krishnamurthy and Sen 2001) and can go out and this entry and exit may happen for a finite number of times. ***Cycling*** in the SA is a counter-example to show that the SA is not polynomial.

For example, the LP (Bazaraa et al. 1990, Beale 1955, Lakshmikantham et al. 2000, Wagner 1969) for which cycling has occurred is

Minimize $z = \mathbf{c}^t\mathbf{x} = [-3/4 \quad 150 \quad -1/50 \quad 6 \quad 0 \quad 0 \quad 0]^t\mathbf{x}$ *subject to* $A\mathbf{x} = \mathbf{b}$, where

$$A = \begin{bmatrix} 1/4 & -60 & -1/25 & 9 & 1 & 0 & 0 \\ 1/2 & -90 & -1/50 & 3 & 0 & 1 & 0 \\ 0 & 0 & 1 & 0 & 0 & 0 & 1 \end{bmatrix}, \quad \mathbf{b} = \begin{bmatrix} 0 \\ 0 \\ 1 \end{bmatrix},$$

where $\mathbf{x} = [x_1 \quad x_2 \quad x_3 \quad x_4 \quad x_5 \quad x_6 \quad x_7\,]^t \geq \mathbf{0}$ (null column vector). The optimal solution is $\mathbf{x} = [1/25 \quad 0 \quad 1 \quad 0 \quad 3/100 \quad 0 \quad 0]^t$, where $z = -1/20$.

Although the unending cycling is an extremely rare phenomenon in the SA, this shows, through the foregoing counter-example, that the SA cannot be even called exponential. The SA could even fail to produce an optimal solution.

LP: deterministic noniterative exponential algorithm From the ***fundamental theorem of linear programming***[7] (Krishnamurthy and Sen 2001), for an LP having k variables and m constraints, there are ${}^kC_m = k!/(m!(k-m)!)$ ways of selecting m of n columns and hence ${}^kC_m$ possible basic solutions.

---

[6]Earlier Khachian's ellipsoid algorithm (Khachian 1979) was an interesting development. Although the ellipsoid algorithm is polynomial-time in integer model, Traub and Wozniakowski have shown that it has unbounded complexity in the real number model (Traub and Wozniakowski 1982).

[7]Consider the LP. *Maximize* $\mathbf{c}^t\mathbf{x}$ *subject to* $A\mathbf{x} = \mathbf{b}$, $\mathbf{x} \geq \mathbf{0}$, where A is $m \times k$ matrix ($k \geq m$) of rank m. If there is a feasible solution then there is a basic feasible solution and if there is an optimal feasible solution then there is an optimal basic feasible solution.

Thus this theorem yields the solution of LPs by searching over a finite number of basic feasible solutions. The complexity of this procedure is combinatorial and hence exponential and so this procedure is not useful in practice for large number of variables. The SA is an improvement over the method of proof of the theorem and the theorem itself. However, in the strict mathematical sense, the SA which, unlike the fundamental theorem based procedure, could encounter failure in extremely rare situation, i.e., in cycling, has ruled in the arena of linear optimisation for over four decades. It is still an algorithm widely used the world over. It may be observed that the Karmarkar method is too expensive for small LPs[8] compared to the SA. For sufficiently large problems, however, the Karmarkar method and other polynomial methods (Barnes1986; Renegar 1988; Vaidya 1990) do excel as these should (because of polynomial complexity). Observe that all the foregoing algorithms are mathematically iterative. For solving LPs, we are yet to have ***polynomial-time noniterative*** algorithms, the development of which is an ***open*** problem.

Specific nonlinear optimisation problems: noniterative polynomial algorithms It may be seen that one of the (usually) infinite solutions of the linear equality constraints or simply linear system $A\mathbf{x} = \mathbf{b}$ will be the solution of the LP (when it has a nonnegative solution), where A is $m \times n$. To get one of these infinite solutions, which could be the minimum-norm least-squares (mnts) solution or a minimum-norm (mn) solution or a least-squares (ts) solution or any other solution noniterative polynomial $O(n^3)$ algorithms (Sen and Prabhu 1976; Sen and Krishnamurthy 1974; Greville 1959; Golub and Kahan 1965; Rao and Mitra 1971; Ben Israel and Greville 1974; Lord et al. 1990; Krishnamurthy and Sen 2001) are available. Observe that the problem of getting the mnts or an mn or a ts solution of $A\mathbf{x} = \mathbf{b}$ is a nonlinear optimisation problem with linear constraints. For the mnts solution (for consistent or inconsistent equations), we minimize $\|\boldsymbol{\delta}\| = \|A\mathbf{x} - \mathbf{b}\| = \sqrt{(\delta_1^2 + \delta_2^2 + \ldots + \delta_m^2)}$ as well as $\|\mathbf{x}\| = \sqrt{(x_1^2 + x_2^2 + \ldots + x_n^2)}$, where $\|\ \|$ denotes the Euclidean norm. For an mn solution (for consistent equations), we minimize $\|\mathbf{x}\|$ while for a ts solution, we minimize $\|A\mathbf{x} - \mathbf{b}\|$.

Integer multiplication: school method versus FFT The widely known school method to multiply two n digit integers needs each digit of the multiplier to be multiplied by each digit of the multiplicand. So the number of basic operations on digits is $O(n^2)$. The sum of two n digit integers, on the other hand, can be computed in O(n) digital operations. Thus multiplication appears harder than addition. Using the fast Fourier transform (FFT) algorithm (Cooley and Tuke 1965; Schilling and Harris 2002) the integer multiplication can be performed in $O(n.\log_2 n)$ operations. If the number of digits

[8] There are numerous practical (real-world) problems in this category.

is n = 32 in each of multiplicand and multiplier then the school method would take $O(n^2)$ = O(1024) basic operations while the FFT would take $O(32.\log_2 32)$ = O(160) basic operations. Thus the FFT needs only about (160 ×100)/1024 = 15.6% of basic operations needed by the school multiplication when n= 32. If n = 64 then the FFT needs only about (384 × 100)/4096 = 9.4% basic operations. This fast multiplication using the FFT was not recorded/known before 1962. For a sufficiently large n, this percentage becomes negligible.

It is not yet known whether the multiplication is harder than addition. We have not yet proved that the multiplication using the FFT is the most optimized way. Maybe, that the multiplication can be performed in O(n) operations. Proving/disproving any of these statements will be a landmark discovery in mathematics, specifically in computational complexity.

3.2.9 Computational intractability: Undesirability of exponential algorithms

To use a computer efficiently, it is necessary to know

(i) how the presentation of an algorithm (program) to a machine (computer) be organized,

(ii) how the efficiencies of two algorithms for the same type of problems be compared,

(iii) if there are problems for which it is impossible to design efficient algorithms, i.e., the problems though solvable by a TM can never be solved due to an excessive amount of computation/time required, and

(iv) if there are procedures/ways to make inefficient (slow) algorithms efficient (fast) introducing a random choice or a guess.

Point (i) is taken care of by sufficiently developed structured programming (single entry single exit modules). It assists in checking whether the program halts, is equivalent to another program, and is correct (Aho et al. 1974).

Point (ii) needs a measure of the complexity of an algorithm. This measure does not depend on the properties of the computer employed for implementing the algorithm. The measure, rather the dynamic complexity measure, that is expressed as a function of the size of the input data needs the specification of the data. One approach is to assume that the input data for a given problem is the worst possible while the other approach is to assume that it is average. The former approach provides us the worst case complexity measure that provides a bound on complexity for practical computation. The later one tells us the average performance of an algorithm.

Point (iii) talks about the problems for which only exponential time, i.e., inefficient algorithms are known. The only algorithms that are considered efficient or fast for general computation are those whose execution time is a polynomial of the input size. ***The input size, for a problem can be taken as the length of the input data in some alphabet.*** Consider, for example, the system of strictly linear inequalities

$$\mathbf{a}_i^t\mathbf{x} < b_i \quad i = 1, 2, \ldots, m,$$

where $\mathbf{a}_i^t \in Z^n$ is the i-th row vector and $b_i \in Z$ is a scalar. Define

$$L = [\Sigma \log_2(|a^t_{ij}| + 1) + \Sigma \log_2(|b_i| + 1) + \log_2 nm + 1,$$

where L = the space needed to state the input of the problem = the number of bits needed to represent the input of the system of inequalities = the length of the input data in binary alphabet with symbols 0 and 1.

If the inequality is $x_1 + x_2 - 4x_3 < -1$ then $m = 1$, $n = 3$, $\mathbf{a}_1^t = [1 \;\; 1 \;\; -4]$, $b_1 = -1$. Hence L can be written, in MATLAB, as

$$L = \log_2(2) + \log_2(2) + \log_2(5) + \log_2(2) + \log_2(3) + 1$$

which is L = 7.9069. Thus, for numerical input, the length of its binary notation, i.e., the number of the symbols 0 and 1 used to represent the input data is used as the input size. If n is the input length then the number of operations in polynomial algorithms is of the form n^k, e.g., n^3. If these are exponential then the number of operations is of the form k^n, e.g., 3^n, where k is a constant.

Evidently the polynomial algorithms are far more efficient than exponential algorithms. Let us have two algorithms A1 and A2 to solve the same type of problems. Let the complexities of A1 and A2 be $O(n^3)$ and $O(3^n)$, respectively and A1 and A2 can be accommodated within the allotted 4 hours of computing time. If the (longest) input length for A2 be g then A2 will perform 3^g steps in 4 hours. Let there be another computer which is 100 times faster. Then for the given problem, we can handle an input of length h such that

$$3^h = 100 \times 3^g$$

or

$$h = g + \log_2 100/\log_2 3 \; = g + 6.6439/1.5850 = g + 4.1917.$$

This implies that we have achieved very little improvement in the performance of the exponential algorithm A2 although the second computer is 100 times faster.

If g is the input length of A1 then we can handle in the faster computer an input of length f, where

$$f^3 = 100 \times g^3$$

or

$$f = (100)^{1/3} \times g = 4.6416 \times g.$$

This performance is certainly better.

If the average time taken for an operation is 50 ns (1 ns = 10^{-9} sec) and the input size is 100 then A1 takes

$$100^3 \times 50 \times 10^{-9} \text{ sec} = 0.05 \text{ sec.}$$

while A2 takes

$$3^{100} \times 50 \times 10^{-9} \text{ sec} = 8.1413 \times 10^{32} \text{ years!}$$

The execution time, viz., 8.1413×10^{32} years, taken by the exponential time algorithm A2 is the time which is very much larger than the estimated age of the universe! Hence, although the problem is solvable by a TM corresponding to A2, it can never be solved due to too much time required for the solution on even an extremely fast computer. On the other hand, the polynomial time algorithm A1 just takes 0.05 sec which is relatively negligible.

A computer which performs 10^8 operations per sec. can execute the O(n), O(n.$\log_2$n), O(n^2), O(n^3), O($n^{3.5}$), and O(n^4) algorithms, where n is the input size, in a reasonable time. It cannot hope to complete O(2^n), O(3^n), and O(10^n) exponential algorithms even if the computer is made several billion times faster. This fact is illustrated in Table 3.2.

Thus even the future computers with much superior technology will have negligible effect on speeding up the exponential algorithms. Hence the polynomial time algorithms are most desirable while the exponential time algorithms are most undesirable. For a very small input size n, the exponential algorithm may be used, but such a small problem is usually of no practical value. A problem is called computationally ***intractable*** if no polynomial time algorithm is available for it.

The intractable problems may be classified into two categories. One category involves decidable or solvable problems in logic, languages, and machines and takes excessive time for the execution. The problems in this category are Turing solvable and decidable. These are called provably intractable since we can prove that even the most unreasonable machine model, e.g., a nondeterministic computer, cannot solve them in polynomial time.

There are, however, certain problems for which a proof on their intractability has yet to be put forth.

Table 3.2: Polynomial versus exponential algorithms: performance measure

Input size n Complexity Function	10	40	70	100
n	1×10^{-7} sec	4×10^{-7} sec	7×10^{-7} sec	1×10^{-6} sec
$n.\log_2 n$	3×10^{-7} sec	2×10^{-6} sec	4×10^{-6} sec	7×10^{-6} sec
n^2	10^{-6} sec	2×10^{-5} sec	5×10^{-5} sec	1×10^{-4} sec
n^3	1×10^{-5} sec	6×10^{-4} sec	.0034 sec	.01 sec
$n^{3.5}$	3×10^{-5} sec	.004 sec	.0287 sec	.1 sec
n^4	10^{-4} sec	.0256 sec	.2401 sec	1 sec
2^n	1×10^{-5} sec	1×10^4 sec	4×10^3 centuries	4×10^{12} centuries
3^n	6×10^{-4} sec	38 centuries	8×10^{15} centuries	2×10^{30} centuries
10^n	100 sec	3×10^{22} centuries	3×10^{52} centuries	3×10^{82} centuries

The other category of intractable (in the strictest sense, i.e., worst case intractability) problems involves all the Turing unsolvable and undecidable problems that no algorithm can solve them.

3.2.10 Turing unsolvable problems

In computer science, there are problems for which the required output will be *yes* or *no*. Consider, for example, the following problems.

(i) ***Fermat's last theorem*** It is easy to write a set of instructions (a computer program) that will search through all positive integers x, y, z, and the integer $n > 2$ for a solution to the equation $x^n + y^n = z^n$. The program will terminate as soon as a solution is obtained. Can we get a solution? In other words, will the program halt?

So far the output (yes or no) has not been found for this problem. It is easy to verify that for $n = 2$, $x = 3$, $y = 4$, $z = 5$, the equation is satisfied.

(ii) ***Infinite loop*** (a) Does a given program (irrespective of any programming language) loop for ever on some input? (b) Does a given program ever produce an output? (c) Does a given program halt on the given input?

(iii) ***Equivalence*** Are the two context-free grammars (of formal languages) equivalent?

(iv) ***Emptiness of context-sensitive grammars*** Is the language generated by a context-sensitive grammar empty?

The following fixed-point iterative (Newton's) scheme to compute a square-root of a real number α correct up to 4 significant digits, viz.,

$$x_{i+1} = 0.5(x_i + (\alpha/x_i)) \quad i = 0, 1, 2, \ldots, \text{till } |x_{i+1} - x_i|/|x_{i+1}| \leq 0.5 \times 10^{-4},$$

where x_0 is an initial approximation, will enter into an infinite loop if α is chosen as a negative number. If we take $\alpha = 9$ and follow the foregoing scheme (algorithm) with $x_0 = -5.5$ then the iterates will be as follows. $x_1 = -3.5682$, $x_2 = -3.0452$, $x_3 = -3.0003$, $x_4 = -3.0000$. If we take $\alpha = -9$ with $x_0 = -5.5$ then the iterates are: $x_1 = -1.9318$, $x_2 = 1.3635$, $x_3 = -2.6186$, $x_4 = 0.4092$, $x_5 = -10.7923$, $x_6 = -4.9792$, $x_7 = -1.5858$, $x_8 = 2.0447$, $x_9 = -1.1784$, and so on (non-terminating). The foregoing scheme works only for nonnegative α and enters into an infinite loop for negative α. Hence the scheme cannot be called an algorithm for negative α. It can at best be called a ***semi-algorithm***. This problem of finding a square-root of a negative number using the foregoing scheme is TM unsolvable (undecidable).

The intractability of a problem does not depend on the encoding scheme employed for inputting the problem and the computational models (Krishnamurthy and Sen 2004; Mishra and Chandrasekaran 2002) such as TM, Push-down stack memory machine (PDM).

3.2.11 NP-class of problems and NP-completeness

Background Myhill (Myhill 1960) studied linear bounded automata and Smullyan (Smullyan 1961) explored rudimentary sets. These models examined space and time bounded machines but did not provide a procedure for complexity measures. Rabin (Rabin 1963) demonstrated that problems can be solved faster on two-tape TMs than on one-tape TMs. Hartmanis and Stearns (Hartmanis and Stearns 1965) quantified time and space complexity on multitape TMs (Fortnow and Homer 2002) and concluded that given more time and/or space, the amount of computation can be increased. Hennie and Stearns (Hennie and Stearns 1966) observed that a two-tape TM can simulate any constant tape machine with a small extra time (a logarithmic factor of time of execution). They gave the time hierarchy, getting a separation if $t_1(n) \log t_1(n) = o(t_2(n))$, i.e., $t_1(n)/t_2(n) \to 0$ as $n \to \infty$.

Savitch (Savitch 1970) proved that problems computable in nondeterministic space s(n) are computable in deterministic space $s^2(n)$. Ibarra (Ibarra 1972) showed that there are problems computable in nondeterministic space n^a but not in space n^b for $a > b \geq 1$. Cook (Cook 1973) proved that problems computable in nondeterministic time n^a but not time n^b for $a > b \geq 1$. Immerman (Immerman 1988) and Szelepcsenyi (Szelepcsenyi 1988) proved independently that nondeterministic space is closed under complement.

NP-problems A problem is assigned to the NP (nondeterministic polynomial time) class if it is verifiable in polynomial time by a nondeterministic TM[9]. A P-problem (whose solution time is bounded by a polynomial) is always also NP. If a problem is known to be NP, and a solution to the problem is somehow known, then demonstrating the correctness of the solution can always be reduced to a single P (polynomial time) verification. If P and NP are *not* equivalent, then the solution of NP-problems requires (in the worst case) an exhaustive search.

Linear program, long known to be NP and thought *not* to be P, was shown to be P by L. Khachian in 1979 in integer model. It was shown to be P by N. Karmarkar in 1984 in real number model. It is an important open problem in computer science to determine if all apparently NP problems are actually P.

NP-completeness The complexity theory was systematically explored since 1960's and started playing a central role in ***computer science*** since 1970's. It found its applications in numerical computation, more generally, in computational science. Several problems of great practical importance resisted polynomial time solution. These problems are mainly optimisation problems such as the Hamiltonian cycle, TSP, certain scheduling problems. In a landmark paper, Karp (1972) showed that 21 intractable combinatorial computational problems are all NP-complete. These problems included the clique problem[10], the set cover problem (an application of this problem is Boolean logic minimization), the independent set problem, and the TSP. S. Cook (Cook 1971) showed that the satisfiability problem[11] is NP-complete while L. Levin (Levin 1973) proved that a variant of tiling problem is NP-complete. A proof of NP-completeness indicates the worst case intractability of a problem. Currently around 300 problems are known to be NP-complete.

---

[9] A nondeterministic TM is a "parallel" TM which can take many computational paths simultaneously, with the restriction that the parallel TM cannot communicate.

[10] Find the largest clique (largest complete subgraph) in an undirected graph.

[11] The boolean satisfiability problem is a decision problem considered in complexity theory. An instance of the problem is defined by a boolean expression written using only AND, OR, NOT, variables, and parentheses. The question is:" Given the expression, is there some assignment of true and false values to the variables that will make the entire expression true?"

Let P = the set of decision problems which can be solved by deterministic TM with a polynomial time bound. Also let NP = the set of decision problems which can be solved by nondeterministic TM with a polynomial time bound. Then P $\subseteq$ NP since the nondeterministic TM can be used as a deterministic TM. Nobody has proved or disproved whether P = NP. However, it is generally believed that P $\neq$ NP. In fact, P = NP problem is one of the seven millennium prize problems (Fortnow and Homer 2002).

If a problem is shown NP-complete, attempts are made to solve it using (i) an approximation algorithm just for a usable solution not far from the true/optimal solution or (ii) a probabilistic algorithm (could be a genetic algorithm or an evolutionary approach) using pseudo-random numbers for a solution in which we have fairly high confidence, say, 95% or 99% (but not 100%).

The set of NP-complete problems is a subset of NP problems. There are NP problems which are not NP-complete. The following diagram (Figure 3.3) depicts that the union of NP-complete problems and P problems is a subset of NP problems

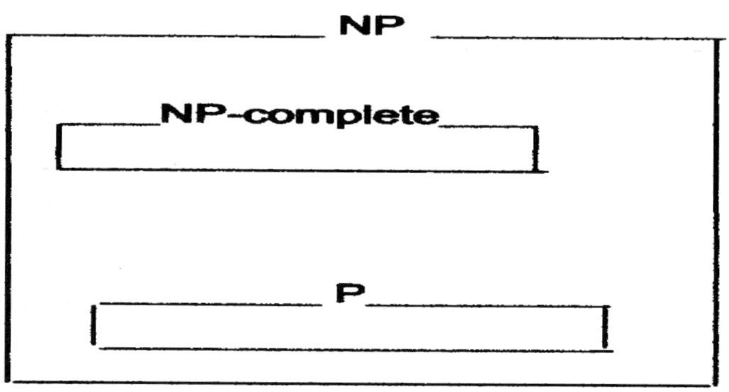

Figure 3.3: NP-complete $\cup$ P $\subset$ NP

Observe that the set of P problems does not belong to that of NP-complete problems. For if it does then we could solve any NP problem in polynomial time. Thus the intersection of P and NP-complete is empty.

Evidently the set of NP-complete problems is smaller than the set of NP problems. Do the NP-complete set and the P set cover the whole NP set? No one has been able to prove that there is a problem in the NP set that is neither in the P set nor in the NP-complete set. However, it has been proved that if P = NP then such a problem must exist. The integer factorisation problem or, equivalently, the composite number problem is such a possible problem.

In the ***integer factorisation problem*** we have to compute two positive integers P > 1 and Q > 1 for a given a positive integer I > 1, such that I = P × Q. It is easy to prove that this problem is in NP just by multiplying all possible two integers ≤ I/2). But so far nobody has proved that it is in P or it is in NP-complete.

NP-hard problems Any decision problem, whether or not in NP, to which we can transform/reduce an NP-complete problem is not solvable in polynomial time unless NP = P. Such a problem is called NP-hard because it is as hard as (or harder than) the NP-complete problems. Observe that every NP-complete problem is an NP-hard problem. The following diagram (Figure 3.4) shows the domain of NP-hard problems among all problems.

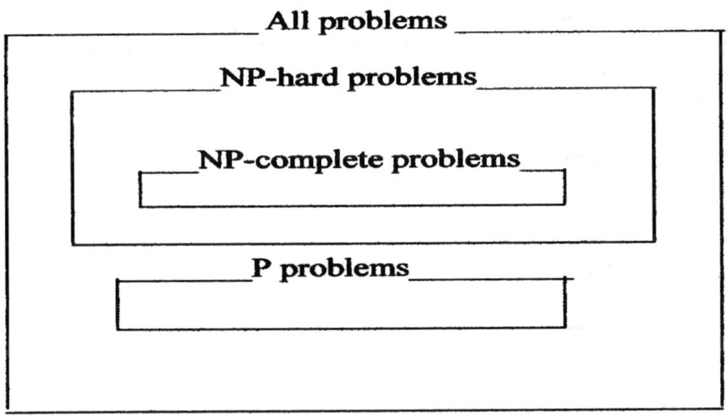

Figure 3.4: Domain of NP-hard problems. All NP-complete problems are NP-hard. P-problems do not belong to NP-hard.

P does not belong to NP-Hard. The intersection of P and NP-hard is empty. This is because if the NP-hard problem can be solved in a polynomial time then according to the definition any NP-complete problem can be solved in polynomial time and thus NP can be solved in polynomial time and thus P = NP.

We **do *not know*** which NP-hard problems are **not** NP-complete. However, if we find such a problem, we would prove that P ≠ NP. Suppose that P = NP, then all NP can be solved in polynomial time and thus all NP-complete problems are solved in polynomial time and hence all problems are NP-hard — a contradiction. The TSP which is an NP-complete problem is an example of NP-hard problem. For an NP-complete problem, (i) the ***solution*** as well as (ii) the ***verification*** of the solution cannot be obtained by polynomial time algorithms.

Handling NP-Hard Problems Just like solving NP-complete problems, there are two approaches. One approach is to develop an approximation algorithm that does not guarantee to give an optimal solution, but rather yields solutions reasonably close to the optimal. The other approach is to develop probabilistic/randomized algorithms. In both the approaches, the algorithms used are polynomial time and the resulting solution, though not guaranteed to be the optimal solution, is reasonably good for practical usage.

An Approximation Algorithm for the Shortest Path Problem (SPP) *The SPP is NP-hard, is a form of TSP, and is stated as follows.*

SPP *Given an undirected graph, where m, n, x denote nodes, w(m, n) the weight between nodes m and n, there is an edge connecting every two nodes, and* $w(m, n) \le w(m, x) + w(x, n)$, *determine the shortest (minimal weight) path starting and ending in the same node and touching all the other nodes only once.*

The steps of the approximation algorithm are as follows.

S1 Obtain a minimum spanning tree.

S2 Create a path that passes twice around the tree.

S3 Change it to the path that does not pass through any node twice.

The path thus obtained is not optimal (shortest) in general. However, this path is reasonably close to the shortest path. In fact, the obtained (computed) path is less than twice the shortest path. In other words, the weight of the computed path is less than twice the weight of the minimal tree.

The SPP (TSP) can also be solved by a probabilistic algorithm, say, the simulated annealing method (Press et al. 1984). This algorithm is polynomial (not exponential). The output, i.e., the computed path is not guaranteed to be the minimal but considered to be pretty close to the minimal path and can be used in a real world situation. We do not yet have a polynomial time algorithm to verify that the computed path is optimal or not. Thus the SPP (TSP) is still NP-complete (NP-hard).

3.3 PSPACE

Kleene (Krishnamurthy and Sen 2004) defined the arithmetic hierarchy. Analogously, Meyer and Stockmeyer (Meyer and Stockmeyer 1972) defined the *polynomial hierarchy* or, equivalently polynomial time hierarchy (PH). This hierarchy is useful in classifying many hard combinatorial/decision problems which do not lie in NP. While most hard decision problems have been shown NP-complete, a few of them have not been classified.

All problems in the polynomial hierarchy are recursive and form a small subset of all recursive problems. There are problems which are recursive and are not captured by the polynomial hierarchy and result in several larger complexity classes that include the PH. One such class of problems is PSPACE. The PH consists of an infinite sequence of classes within PSPACE. A problem in PSPACE can be solved using storage/work space which is of polynomial length relative to the input size of the problem.

The zeroth (bottom-most) level of hierarchy is class P. The first level of hierarchy is class NP. The second level of hierarchy are all problems in NP relative to an NP oracle[12]. Iterating this idea to all finite levels produces the complete hierarchy. It is believed that each level of the PH is a proper subset of the next higher level. If P = PH then the complete PH reduces to class P. In fact, P ≠ PH is widely believed. While every class in the PH is contained in PSPACE, the converse is false if the hierarchy is strict.

The PSPACE-completeness is defined in the same way as the NP-completeness. Checkers played on boards of finite (but unbounded) size is a PSPACE problem. In fact, several PSPACE problems are found in generalized games (Garey and Johnson 1979). The exponential time and exponential complexity classes exist beyond PSPACE.

3.4 Alternation

The alternation deals with the classification of combinatorial problems using alternating TM — a generalization of nondeterministic TM. For the work on alternation, the reader may refer Chandra et al. (1981).

---

[12] According to Cook (1971), a problem A in NP is NP-complete if, for every problem A′ in NP, there is a polynomial algorithm in which the answer to questions like "what is the answer to the input I with respect to problem A?" can be included and used. Cook calls such a question-answering device an ***oracle***. It looks like a subroutine. If there is a polynomial algorithm for some NP-complete problem A then P = NP in Cook's definition. Each call of the oracle uses a polynomial time to answer and there can only be a polynomial number of such calls that make the solution process polynomially bounded.

3.5 LOGSPACE

All the complexity classes considered so far consist of the class P of the polynomial time computable problems. In class P, there are problems for which smaller space classes, viz., deterministic log space class L and nondeterministic log space class NL may be considered. It may be checked that L $\subseteq$ NL $\subseteq$ P. For graph connectivity as well as word problems, logspace algorithms have been developed (Feige 1996; Lipton and Zalcstein 1977; Fortnow and Homer 2002).

3.6 Probabilistic complexity

There are many important real world problems which are branded NP-complete. Since they need to be solved, we try to design some usable algorithms for them. As already mentioned, two different approaches, viz., an approximative approach (approximation algorithms) and the probabilistic approach (probabilistic algorithms) to cope with such problems. The probabilistic algorithms (Hammersley and Handscomb 1965; Gordon 1970), e.g., the Monte Carlo methods make use of random choices and have been in use for long. It has been shown (Rabin 1976) more recently that the probabilistic algorithms can solve some NP-complete problems more efficiently (in terms of time and space complexity) than the known deterministic algorithms. In fact, Rabin (1976) and Strassen and Solovay (Strassen and Solovay 1977) designed probabilistic algorithms for testing whether a number is prime in polynomial time with a small probability of error. Observe that the primality problem and its complement (composite number problem) are in the NP-class. However, recently Agrawal et al. (2002) gave a deterministic polynomial algorithm for primality. Their design suggests that the probabilistic algorithms may be useful for solving other deterministically intractable (because of excessive time needed for computation) problems.

The Strassen and Solovay probabilistic algorithm generated random numbers (flipped coins) to help search for a counterexample to primality. The algorithm was based on the argument that if the number was not prime then with a very high probability a counterexample could be found. All the probabilistic algorithms are usually meant to solve problems in the NP-class (including NP-complete) and are polynomial time. The outputs of these algorithms, unlike those of the corresponding deterministic algorithms, cannot be always said correct with 100% confidence. Yet these are usable in the real world situation. Thus, for probabilistic algorithms, we produce the results (outputs) with which we attach the confidence level (say, 95% or 99% and never 100%).

Gill (1977) has studied thoroughly the complexity of probabilistic TMs and developed a valuable model for probabilistic algorithms with built-in random decisions. His study led to the following conjectures.

Conjecture 1 There is a function computable probabilistically in polynomial time but not computable deterministically in polynomial time.

Conjecture 2 There is a function computable probabilistically with bounded error probability in polynomial time but not computable deterministically in polynomial time.

Conjecture 3 There is a function computable probabilistically in polynomial bounded average running time but not computable deterministically in polynomial time.

Rabin (1976) and Strassen and Solovay (1977) showed that a prime can be recognized in polynomial time with bounded error probability and thus supported Conjecture 2. However, these conjectures appear to contradict the well-known theorem (Manna 1974) that *the class of nondeterministic TMs has the same computing power as that of deterministic TMs.* But this is not so because Turing's concept of computing power, as pointed out earlier, is based not on the complexity measure but on the inherent solvability (decidability) of a given problem.

Besides the primality detection, the probabilistic algorithms can be used to prove the correctness of a program, which can be shown by constructing suitable *witnesses* for incorrectness using different test inputs. A few randomly chosen test inputs will ensure provably high probability of correctness. These can also be used to solve TSP and to find the Hamilton path in a graph.

Not only for NP-class problems, but also for some problems for which deterministic polynomial algorithms are available, the probabilistic algorithms could be used rather more conveniently and beneficially. Consider, for example, the numerical single or multiple integration problems (Krishnamurthy and Sen 2001). These problems can be solved using the deterministic polynomial algorithms such as the Simpson's 1/3 rule (close quadrature formula), Gauss-Legendre (open) quadrature in polynomial time. These also can be solved using the Monte Carlo techniques (Krishnamurthy and Sen 2001) which can be more easily programmed and which would perform better in some complicated multi-variable functions. These Monte Carlo algorithms will ensure provably 100% probability (implying 100% confidence) of correctness with error bounds like those obtained in closed/open quadrature formulas.

3.6.1 Interactive proof systems

The notion of proof system can be generalized by permitting probabilistic verification of the proof. Interaction can be considered when the verifier sends messages based on flipping random coins. Babai (1985) defined interactive proof system for classification of some group questions. Goldwasser et al. (1989) defined an alternative interactive proof system, called the zero-knowledge proof system, for the cryptographic class zero-knowledge.

3.7 Descriptive complexity

Descriptive complexity attempts to measure the computational complexity of a problem in terms of the complexity of the logical language required to state the problem. Fagin (1973) was the first to give a theorem which states that NP is exactly the class of problems definable by existential second order Boolean formulas and which gives a logical characterization of the NP-class.

3.8 Boolean circuit complexity

A Boolean circuit is a directed acyclic graph whose internal nodes (gates) are Boolean functions such as *and, or, not.* A circuit with k inputs may be considered as a recognizer of a set of strings each of length k, viz., those which led to the circuit evaluating to 1. For further information, refer Fortnow and Homer (2002) and Razborov (1985).

3.9 Communication complexity

The communication complexity aims at modelling the efficiency and complexity of communication between computers. However, intra-computer communication, for example, the communication between a processor and the executable shared memory or between the cache and the executable memory or the cache and a processor (all belonging to one computer) is also important and the concerned complexity is also studied. The communication complexity determines how much data/information need to be exchanged between two computers to carry out a given computation with the assumption that both the computers have unlimited computational power. For further details, refer Kushilevitz and Nisan (1996).

3.10 Quantum complexity

Recently to analyze the computational power of quantum computers (no commercial quantum computers are so far existing), the quantum complexity is studied. R. Feynman (1982) observed that the conventional computers based on silicon technology could not efficiently simulate the quantum systems. He felt that if a computer could be built based on quantum mechanics then it might be able to perform the task more efficiently. Such a theoretical computational model was developed by D. Deutch (1985). Two quantum algorithms (Shor 1997; Grover 1996) received significant attention. One algorithm was for factoring an integer in a polynomial time on a quantum machine while the other for searching a database of n elements in $O(\sqrt{n})$ operations/time.

3.11 Parallel complexity

Complexity for parallel/overlapped computing is another important area which has been significantly studied. For this one can consider a general configuration of a parallel computer with different levels of parallelism or a specifc computing model (Quinn 1987; Schendel 1984). Figure 1.2 of Chapter1 depicts a general configuration of a computer. A general parallel computer (Figure 3.5) may be diagrammatically represented as

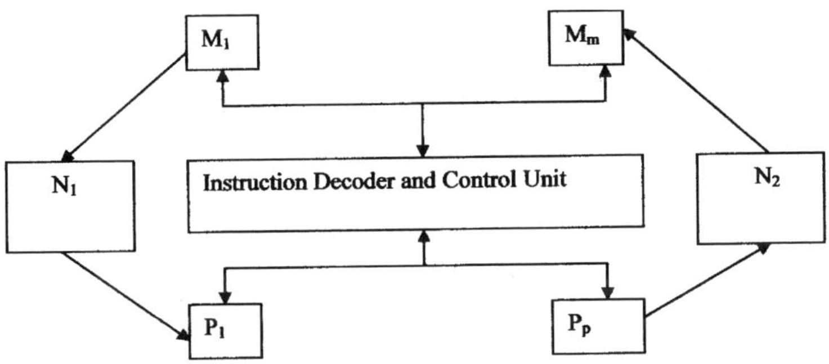

Figure 3.5: General parallel computer configuration

M_i = Memories P_i = Processors, N_i = Control and Data Organization Networks

The parallelism could exist (i) within the control unit, (ii) among the processors P_i, (iii) among the memories M_i, and (iv) in the networks N_i.

Computing devices have been classifieded by Flynn (1966) based on number of data and instruction streams. These are SISD (classical von Neumann), SIMD (includes array processors and pipeline processors), MISD (chains of processors and is equivalent to SISD and hence is not so important), and MIMD (multiple processor version of SIMD) models, where SI = single instruction stream, SD = single data stream, MI = multiple instruction stream, MD = multiple data stream (Quinn 1987). Keeping in view the computing model that is used, we may define the ***speed-up ratio*** S. However, it is important in parallel computation to be able to assess, irrespective of any specific parallel model, the speed gain expected from the operation of p processors P_i in parallel. For this, the ratio S_p for an algorithm is defined as $S_p = T_1/T_p \geq 1$, where T_1 = computing time on a sequential computer and T_p = computing time on a parallel computer with p processors. If $k < 1$ is a positive fraction close to 1 and is a computer dependent parameter then the speed-up ratio S_p of the parallel computer has the forms (i) $S_p = kp$ (matrix computation), (ii) $S_p = k \log_2 p$ (searching), (iii) $S_p = kp/\log_2 p$ (linear tridiagonal equations, linear recurrences, sorting, polynomial evaluation), (iv) $S_p = k$ (compiler operations, nonlinear recurrences). The efficiency (utilization) of the parallel machine (algorithm) is then defined as $E_p = S_p/p \leq 1$.

The measure of effectiveness $F_p = T_1/(pT_p^2)$ of a parallel machine may be used to compare two parallel algorithms for the same problem. It can be seen that $F_p = E_pS_p/T_1 \leq 1$ depends on both speed up ratio and efficiency. Since the efficiency E_p is directly proportional to S_p for a fixed number of processors p, the effectiveness F_p is directly proportional to S_p^2 assuming T_1 constant. Thus finally it is the square of the speed up ratio S_p that needs to be maximized for the best performance of the parallel machine/algorithm. The performance measure may be defined as $R_p = F_p \times T_1$. We may compute R_p of the parallel algorithm for a given problem on a machine with different number of processors.

Consider, as an example, the multiplication of 32 numbers c_i, i.e.,

$$\text{Product} = \Pi c_i \quad (i = 1(1)32)$$

A single processor machine will need 31 multiplications. If we assume that 1 multiplication is done in 1 unit time then we have $T_1 = 31$. A two processor machine would compute Product1 on processor P_1 and Product2 on processor P_2 simultaneously as

$$\text{Product1} = \Pi c_i \quad (i = 1(1)16),$$
$$\text{Product2} = \Pi c_i \quad (i = 17(1)32)$$

needing only 15 time units and then 1 time unit to get Product = Product1 × Product2 at the next stage. Thus the two processor machine would need T_2 = 16 time units.

If we have a three processor machine then we compute Product1 on processor P_1, Product2 on processor P_2, and Product3 on processor P_3 simultaneously as

$$\text{Product1} = \Pi c_i \quad (i = 1(1)11), \text{(10 time units)}$$
$$\text{Product2} = \Pi c_i \quad (i = 12(1)22), \text{(10 time units)}$$
$$\text{Product3} = \Pi c_i \quad (i = 23(1)32), \text{(nine time units)}$$

needing only 10 time units. We then need 2 time units on processor P_1 to compute Product4 = Product1 × Product 2 and Product5 = Product4 × Product3. Thus we need T_3 = 12 time units in a three processor machine. Given four processors, we would be needing $T_4 = 7 +1 +1 = 9$ time units. For eight processors, it would be $T_8 = 3 + 1 +1 +1 = 6$ time units. This example is a simplistic one as we have not considered the communication overhead (which is significant) among processors. However, a table (Table 3) of performance measures R_p for different number of processors p is as follows.

Table 3.3 Performance measures R_p for different number of processors

p	T_p	$S_p=T_1/T_p$	$R_p=T_1^2/(pT_p^2)$
1 (Serial)	31	1	1
2	16	1.94	1.88
3	12	2.58	2.22
4	9	3.44	2.97
8	6	5.17	3.34
16	5	6.20	2.40

Any parallel machine will have two or more processors. One may seek time and processor bounds for a given algorithm run on a parallel machine. We will discuss parallel complexity in more detail in a subsequent chapter.

Bibliography

Agrawal, M.; Kayal, N.; Saxena, N. (2002): *PRIMES is in P*, Unpublished manuscript, Indian Institute of Technology, Kanpur.

Aho, A. V.; Hopcroft, J. E.; Ullman, J. D. (1974): *The Design and Analysis of Computer Algorithms*, Addison-Wesley, Reading, Massachusetts.

Babai, L. (1985): Trading group theory for randomness, in *Proc. 17th ACM Symp. On Theory of Computing,* 421-29, ACM, New York.

Barnes, E. R. (1986): A variation of Karmarkar's algorithm for solving linear programming problems, *Math. Program.*, **36**, 174-82.

Bazaraa, M. S.; Jarvis, J. J.; Sheraldi, H. D. (1990): *Linear Programming and Network Flows*, 2nd ed. Wiley, Singapore, 165-67.

Beale, E. M. L. (1955): Cycling in the dual simplex algorithm, *Naval Research Logistics Quarterly*, **2**, 269-75.

Ben–Israel, A.; Greville, T.N.E. (1974): *Generalized Inverses: Theory and Applications*, Wiley, New York.

Chandra, A.; Kozen, D.; Stockmeyer, L. (1981): Alternation, *J. ACM*, **28**, 114-33.

Cooley, J.W.; Tuke, R.W. (1965): An algorithm for machine computation of complex Fourier series, *Mathematics of Computation*, **19**, 297-301.

Cook, S. (1971): The complexity of theorem proving procedures, In *Proc. 3rd ACM Symp. Theory of Computing*, 151-58.

Cook, S. (1973): A hierarchy for nondeterministic time complexity, *Journal of Computer and System Sciences*, **7**, 4, 343-53.

Dantzig, G. B. (1963): *Linear Programming and Extensions*, Princeton University Press, Princeton, New Jersey.

Davis, M. (1958): *Computability and Unsolvability*, McGraw-Hill, New York.

Deutsch, D. (1985): Quantum theory, the Church-Turing principle and the universal quantum computer, *Proc. Royal Soc. of London A*, 400:97.

Donovan, J. J. (1972): *Systems Programming*, McGraw-Hill, New York.

Fagin, R. (1973): *Contributions to the model theory of finite structures*, Ph.D. Thesis, University of California, Berkeley.

Feige, U. (1996): A fast randomized LOGSPACE algorithm for graph connectivity, *Theoretical Computer Science*, **169,** 2, 147--60.

Felder, K. (1996): Kenny's overview of Hofstadter's Explanation of Godel's theorem, the website http//www.ncsu.edu/felder-public/Kenny/papers/godel.html

Feynman, R. (1982): Simulating physics with computers, *International J. Theoretical Physics*, **21**, 467.

Fortnow, L.; Homer, S. (2002): A short history of computational complexity, the website http://www.neci.nj.nec.com/homepage/fortnow, also the website http://www.cs.bu.edu/faculty/homer.

Garey, M.; Johnson, D. (1979): *Computers and Intractability: A Guide to the Theory of NP-completeness*, W.H. Freeman, San Francisco.

Gass, S. I. (1969): *Linear Programming* (3rd ed.), McGraw-Hill, New York.

Gass, S.I. (1975): *Linear Programming: Methods and Applications*, McGraw-Hill, New York.

Gill, J. (1977): Computational complexity of probabilistic Turing machines, *SIAM J. Comput.*, **6**, 675-95.

Glenn, J. (1996): *Scientific Genius: The Twenty Greatest Minds*, Random House Value Publishing, New York.

Godel, K. (1931): Uber formal unedtscheidhare Satze der Principia Mathematica and verwandter Systeme, I, *Monatshefte fur Mathematik und Physik*, **38**, 173-98.

Godel, K. (1961): *The Consistency of the Axiom of Choice and of the Generalized Continuum-hypothesis with the Axioms of Set Theory*, Princeton University Press, Princeton.

Goldwasser, S.; Kilian, J.; Rackoff, C. (1989): The knowledge complexity of interactive proof-systems, *SIAM J. Comput.*, **18**, 1, 186-208.

Golub, G.; Kahan, W. (1965): Calculating the singular values and the pseudo-inverse of a matrix, *SIAM J. Numer. Anal.*, **B-2**, 205-24.

Gordon, R. (1970): On Monte Carlo algebra, *J. Appl. Prob.*, **7**, 373-87.

Greville, T.N.E. (1959): The pseudo-inverse of a rectangular or singular matrix and its application to the solution of linear equations, *SIAM Rev.*, **1**, 38-43.

Grover, L. (1996): A fast quantum mechanical algorithm for database search, *Proc. 28th ACM Symp. On Theory of Computing*, 212-219, ACM, New York.

Hammersley, J.M.; Handscomb, D.C. (1965): *Monte Carlo Methods*, Methuen, London.

Harary, F. (1972): *Graph Theory,* Addison-Wesley, Reading, Massachusetts.

Hartmanis, J. (1994): On computational complexity and the nature of computer science, *Comm. ACM*, 37(10), 37-43.

Hartmanis, J.; Stearns, R (1965): On the computational complexity of algorithms, *Trans. Amer. Math. Soc.*, **117**, 285-306.

Hennie, F.; Stearns, R. (1966): Two-tape simulation of multi-tape Turing machines, *J. ACM*, **13**(4), 533-46.

Ibarra, O. (1972): A note concerning nondeterministic tape complexities, *J. ACM*, **19**, 4, 608-12.

Immerman, N. (1988): Nondeterministic space is closed under complementation, *SIAM J. Computing*, **17**, 5, 935-38.

Karmarkar, N. (1984): A new polynomial-time algorithm in linear programming, *Combinatorics,* **4**, 373-395.

Karp, R. (1972): Reducibility among combinatorial problems, In *Complexity of Computer Computations*, 85-104, Plenum Press, New York.

Khachian, L.G. (1979): A polynomial algorithm in linear programming, *Dokl. Akad. Nauk USSR*, **244**, 1093-1096, translated as *Soviet Math. Dokl.* **20**, 191-194.

Krishnamurthy, E.V.; Sen, S.K. (2001): *Numerical Algorithms: Computations in Science and Engineering*, Affiliated East-West Press, New Delhi.

Krishnamurthy, E.V.; Sen, S.K. (2004): *Introductory Theory of Computer Science*, Affiliated East-West Press, New Delhi.

Kushilevitz, E.; Nisan, N. (1996): *Communication Complexity*, Cambridge University Press, Cambridge.

Lakshmikantham, V.; Sen, S.K.; Jain, M.K.; Ramful, A. (2000): $O(n^3)$ noniterative heuristic algorithm for linear programs with error-free implementation, *Applied Mathematics and Computation*, **110**, 53-81.

Levin, L. (1973): Universal sorting Problems, *Problems of Information Transmission*, 9, 265-66.

Lipton, R.; Zalcstein, E. (1977): Word problems solvable in logspace, *J. ACM*, 3, 522-26.

Lord, E.A.; Venkaiah, V. Ch.; Sen, S.K. (1990): A concise to solve under-/over-determined linear systems, *Simulation*, 54, 239-240.

Luenberger, D.G. (1973): *Introduction to Linear and Nonlinear Programming,* Addison-Wesley, Reading, Massachusetts.

Manna, Z. (1974): *Mathematical Theory of Computation*, McGraw-Hill, New York.

Meyer, A.; Stockmeyer, L. (1972): The equivalence problem for regular expressions with squaring requires exponential space, in *Proc. Of the 13th IEEE Symposium on Switching and Automata Theory*, 125-29, Massachusetts Avenue, N.W., Washington, D.C., 20036-1903, Conputer Society Press of IEEE.

Mishra, K.P.L.; Chandrasekaran, N. (2002): *Theory of Computer Science: Automata, Languages and Computation* (2nd ed), Prentice-Hall of India, New Delhi.

Myhill, J.(1960): Linear bounded automata, Tech. Note 60-165, Wright-Patterson Air Force Base, Wright Air Development Division, Ohio.

Nagel, E.; Newman, J.R. (1964): *Godel's Proof*, New York University Press, New York.

Press, W.H.; Teukolsky, S.A.; Vetterling, W.T.; Flannery, B.P. (1984): *Numerical Recipes in C/Fortran*, Prentice-Hall of India, New Delhi.

Quinn, M.J. (1987): *Designing Efficient Algorithms for Parallel Computers*, McGraw-Hill, New York.

Rabin, M. (1963): Real time computation, *Israel Journal of Mathematics*, 1, 203-11.

Rabin, M.O. (1976): Probabilistic Algorithms, in Algorithms and Complexity, ed. J.F. Traub, Academic Press, New York.

Rao, C.R.; Mitra, S.K. (1971): *Generalized Inverse of Matrices and Its Applications*, Wiley, New York.

Razborov, A. (1985): Lower bound on the monotone complexity of some Boolean functions, *Doklady Academii Nauk SSSR*, 281, 4, 798-801.

Renegar, J. (1988): A polynomial-time algorithm based on Newton's method for linear programming, *Math. Program.*, **40**, 59-93.

Savitch, W. (1970): Relationship between deterministic and nondeterministic tape classes, *Journal of Computer and System Sciences*, 4, 177-92.

Schendel, U. (1984): *Introduction to Numerical Methods for Parallel Computers,* Ellis Horwood, Chichester

Sen, S.K. (2003): Error and computational complexity in engineering, ed. J. C. Misra, Narosa Publishing House, New Delhi.

Sen, S.K.; Krishnamurthy, E.V. (1974): Rank-augmented *LU*-algorithm for computing generalized matrix inverses, *IEEE Trans. Computers*, C-23, 199-201.

Sen, S.K.; Prabhu, S.S. (1976): Optimal iterative schemes for computing Moore-Penrose matrix inverse, *Int. J. Systems Sci.,***8**, 748-53.

Schilling, R.J.; Harris, S.L. (2002): *Applied Numerical Methods for Engineers using MATLAB and C*, Thomson Asia, Singapore.

Shor, P. (1997): Polynomial-time algorithms for prime factorisation and discrete logarithms on a quantum computer, *SIAM J. Comput.*, **26**, 5, 1484-1509.

Smullyan, R. (1961): *Theory of Formal Systems*, Vol.**47** of Annals of Mathematical Studies, Princeton University Press.

Stearns, R. (1994): It's time to reconsider time, *Comm. ACM*, **37**(11), 95-99.

Strassen, V.; Solovay, R. (1977): A fast Monte Carlo test for primality, *SIAM J. Comput.,* **6**, 84-85.

Szelepcsenyi, R. (1988): The method of forced enumeration for nondeterministic automata, *Acta Informatica*, **26**, 279-84.

Traub, J.F.; Wozniakowski, H. (1982): Complexity of linear programming, *Operations Research Letters*, **1**, No. 1, 59-62.

Vaidya, P. M. (1990): Algorithm for linear programming which requires $O(((m+n)n^2 + (m+n)^{1.5}n)L)$ arithmetic operations, *Proc. ACM Annual symposium on Theory of Computing* (1987), 29-38; Math. Program., **47**, 1990, 175-201.

Vajda, S. (1975): *Problems in Linear and Nonlinear Programming,* Charles Griffin, London.

Wagner, H. M.(1969): *Principles of Operations Research*, 2nd ed., Prentice-Hall, Englewood-Cliffs, New Jersey.

Whitehead, A.N.; Russell, B; (1910-13): *Principia Mathematica*, **1** (1910), **2** (1912), **3** (1913), Cambridge University Press, London.

Winston, W.L. (1994): *Operations Research: Applications and Algorithms*, Duxbury Press, Belmont, California.

Chapter 4

Errors and Approximations in Digital Computers

4.1 Introduction

In the numerical solution of problems in algebra and analysis, the properties of digital computers, which are relevant to their use are:

(i) Computers use only a simulation of real number system, called the floating-point number system and not the real number system. In the floating-point number system, a number is expressed as a fraction (or an integer) and an exponent. This introduces the problem of rounding errors.

(ii) The solution of very large problems is possible due to speed of computer processing. Often large problems have solutions which are much more sensitive to the perturbations of the data than are those of small problems.

(iii) The speed also permits many more operations to be performed in a short time. Consequently, the instability of many algorithms is clearly revealed.

(iv) Since the intermediate results of a computation are hidden in the storage of the computer, it is necessary to ensure that the computation does not fail in an intermediate step.

These properties of digital computers cause many pitfalls such as errors, instability, and obscurities (Forsythe 1970).

This chapter is mainly for those who are deeply involved in large-scale scientific and engineering computations. A clear understanding of what is going on inside the computer helps in debugging as well as minimizing error and reducing the complexity (computational cost). Even for computer scientists who have something to do with numerical computations, this chapter is informative.

4.1.1 What is computation

The execution of instructions/commands by a computer for specified data (numbers, alphanumerical and special characters which are all in 0-1 form) is computation. The word computer literally means any machine (m/c) capable of arithmetic computations, viz., add, subtract, multiply, and divide operations. However, the wider meaning of the word is any m/c with an internal memory[1] that is (i) electronic[2] and (ii) capable of changing the course of execution[3] of instructions as well as, of course, the foregoing arithmetic operations, and also logical (such as AND, OR, and NOT) and character string processing operations (such as comparison, concatenation, insertion, and deletion). Primarily error is introduced in the arithmetic computation while nonarithmetic computation is usually error-free.

4.1.2 Analog computer versus digital computer

There are two main classes of computers — analog and digital. The computer that measures numerically continuous physical quantities such as electrical current, voltage, temperature, pressure, and length, and then realizes an arithmetic or a logical operation (such as AND, OR, and NOT) is called an analog computer. An analog computer that realises a divide operation ($i = v/r$) can be just a circuit with a current source, an ammeter A (that measures current i in ampere) and a variable resistor R (indicating the value of the resistance r in ohms), in series and a voltmeter V in parallel (i.e., across R). A digital computer, often referred to as simply a *computer* defined in Section 4.1.1, on the other hand, operates directly on digits that represent either discrete data or symbols.

---

[1]that can store an ordered set of instructions called *program* and input information called *data* required by the program

[2]The m/c produces the results through the movement of electronic pulses and not by the physical movement of the internal parts.

[3]The m/c while executing the instructions in the program in a sequence changes the course of execution of instructions due to a decision based on data stored in its internal storage or on the outcome of an arithmetic/logical operation where the outcome of a logical operation is *true* or *false*.

The analog computer produces an output in a computation with a high (higher-order) error, i.e., with an accuracy usually not greater than 0.005%[4]. This is because of the inherent error existing in the device measuring a physical quantity. The digital computer has a low (lower-order) error, i.e., it can produce a desired accuracy, say 10^{-13}% or greater than 10^{-13}% in computation subject to the availability of sufficient hardware resources and appropriate software programs. Input data (obtained from the real-world measurement) to a digital computer, however, may not be usually more accurate than 0.005%. A general purpose digital computer can solve a wide range of problems and more versatile than the analog **m/c. Although sometimes the analog computer may produce outputs, say those of an FFT (fast Fourier transform) algorithm for a very large set of complex data points faster than a digital computer, the digital computer is almost always much more accurate in computation.**

4.1.3 Analog input-output error versus digital input-output error

Observe that the analog computer takes exact real-world quantities (which can never be, in general, exactly captured by us) as its input and produces exact quantities as output that can never be, in general, recorded by us due to error present in any measuring device. The digital computer, on the other hand, takes erroneous (due to uncertainty in measurement) input with an error not usually less than 0.005% and computes digitally an output that involves both input error and computational error.

In most cases, however, it is the digital computer that has both the enormous speed (say, 10 billion flops (floating-point operations/sec)) advantage as well as the computational accuracy (much more than 0.005%) advantage. It can be seen that the input data, when obtained as an outcome of a measuring device, for digital computing will not be usually more accurate than 0.005%. Assuming these input data error-free, the digital computer will provide usually a much higher accuracy than that produced by the corresponding analog device. Thus, in almost all real-world situations, the word *computer* will imply only the digital computer and not the analog one. In our discussion throughout this chapter, we will be only concerned with the approximations (i.e., errors in number representations) and computational errors in a digital computer.

---

[4] Observe that a greater accuracy is a relative value less than 0.005%. Thus, an accuracy of 0.001% is greater than an accuracy of 0.005%.

4.2 Number representation

4.2.1 Numerals

To represent a quantity, we prefer certain special symbols with their meanings and hierarchy (order) associated with them. These special symbols are termed as *numerals* or numerical characters. A sequence of these characters represents a number. For example, Roman numerals (such as vi meaning six) and decimal numbers (such as 6 meaning six) constitute two different number systems. The Roman number system is rarely used because of the problems of representing large numbers and performing arithmetic operations on these numbers. The decimal number system, on the other hand, is the most widely used and most widely understood system for representing numbers as well as performing arithmetic operations on them so far as the man-man communication is concerned.

4.2.2 Why decimal number system in man-man communication

We need at least two symbols to represent any number/information. Observe that the blank is a valid symbol. If we introduce the constraints, such as

- (i) the common human *psychology* (around 7 ± 2 things can be comprehended at a time),
- (ii) the *physical length*, i.e., the number of symbol (should be minimal) required to represent an information,
- (iii) the wide-spread (world-wide) *familiarity* of the symbols and their usage in information representation, and
- (iv) the man-man (including man-self) *communication*

then possibly the set of 10 decimal symbols 0, 1, 2, 3, 4, 5, 6, 7, 8, and 9 to represent a quantity in a specified unit is optimal while the 26 Latin alphabets a, b, c, . . ., z along with several other special symbols (many of which are available on a computer keyboard) could perhaps be considered optimal from the man-man or man-self communication point of view.

4.2.3 Decimal symbols differ from language to language

The foregoing ten decimal symbols are usually (not always though) different in different languages in the world. Notably, one extreme is the representation of these symbols in different Indian languages (around 179). These decimal symbols in Kannada or in Hindi or in Tamil are significantly different from those in Bengali or Oriya. The other extreme is the representation of these symbols in European languages in which almost always the foregoing symbols 0, 1, 2, 3, 4, 5, 6, 7, 8, and 9 are used with the same customary meaning and hierarchy attached to each of the symbols. However, in our subsequent discussion we will use only the symbols 0, 1, 2, . . ., 9 with their usual conventional meaning. In a binary number system, we use the two symbols 0 and 1, where 1 succeeds 0 and is greater than 0. In octal (base 8) number system, for example, we use the symbols 0, 1, 2, 3, 4, 5, 6, and 7, where 7 > 6 > 5 > . . . > 0.

4.2.4 Other number systems

While a computer programmer could use entirely the ordinary decimal number system while writing a program, it would be more convenient for him if he knows other number systems, notably binary, octal, and hexadecimal. That is, a knowledge of what is actually going on inside a computer would be more helpful.

Besides the positional number systems in various bases such as 2, 8, 10, and 16 to represent a quantity, and Roman numerals, we have negative radix presentation, variable radix number system, residue number system with multiple prime bases, the p-adic number system with a prime-power base, and several other notations (Krishnamurthy 1971a; Krishnamurthy et al. 1975a, 1975b; Gregory and Krishnamurthy 1984; Sankar et al. 1973a, 1973b).

4.2.5 Binary versus decimal numbers: Physical size

The silicon technology based (hardware) computer has only two states which are stable[5] fast-switching (of the order of nanoseconds)[6]. Consequently all the information in a digital computer is represented using only the two symbols corresponding to two stable physical states (two specific/distinct voltage levels, say). We call these two symbols 0 and 1. All the computations (arithmetic and nonarithmetic) on number/information are carried out on these two symbols. In fact, in the domain of electronic digital computers, *we have so far not found three or more stable fast-switching states*. If at all we have, say, 10 stable fast-switching states then binary number system will loose significantly its importance in the realm of digital computers, and possibly in computer science.

In addition, the physical size of the information will be possibly reduced by about 333% inside the computer storage such as the main memory, CPU registers, the cache, and hard disks.

---

[5] Stability implies that a state continues to remain as it is theoretically for ever until it is changed by some electronic means.

[6] Fast change of one sequence of binary digits to another

4.2.6 Why base 2 system

In nature, we are yet to find more than two truly such stable fast-switching states. Thus, the base 2 system of representation in a computer has been existing for over five decades and possibly will exist for ever.

4.2.7 Positional number systems

Thus, the base 2 system of Thus, the base 2 system of A binary number is a polynomial $\Sigma d_n 2^n$, where the summation is over n = – k (1) s with d_n = 0 or 1 for a particular value of n. The numbers k and s are both nonnegative (usually positive) integers. A number system involving the positive integer base, say, 10, where each digit represents a value by virtue of its position is called the positional number system. A conventional decimal number, say 639 or a conventional binary number, say 11100101 belongs to this positional number system. In the foregoing decimal number 639, the digit 6 represents the value $6 \times 10^2 = 600$, the digit 3 represents the value $3 \times 10^1 = 30$, while the digit 9 represents the value $9 \times 10^0 = 9$. The decimal number 639 is just the sum of these three values. Similarly, the leftmost bit (binary digit) 1 represents the value $1 \times 2^7 = 128$ (in decimal), the next (second from left) digit 1 represents the value $1 \times 2^6 = 64$ (in decimal), and so on. The binary number 11100101 is just the sum of the eight values which is 229 (in decimal). In the same way the octal number 417 (may be in some context represented as $(417)_8$) represents the value $4 \times 8^2 + 1 \times 8^1 + 7 \times 8^0 = 271$ (in decimal).

4.2.7.1 Set of symbols in base β: symbol β not in the set

Observe that the symbol β = 8 does not belong to the octal number system just as the symbol β = 2 does not belong to binary system. However, unlike the single-symbol bases (radices) 2 and 8, we are not used to use a single symbol base (say, A) for the decimal number system; we use two-symbol radix, viz., 10 for the system in which both symbols 0 and 1 of the base 10 are present as two distinct symbols out of the ten symbols. This usage has not/will not usually confuse a reader. Similarly for hexadecimal number system, the base is not used as a single symbol, say, G but as two symbols, viz., 1 and 6; both are individually present among the sixteen symbols 0, 1, 2, 3, 4, 5, 6, 7, 8, 9, A, B, C, D, and F used in the hexadecimal system.

Since people are accustomed to decimal number system, we have indicated the value of the binary number as well as that of octal number in decimal forms which can be easily appreciated/gauged by common human beings.

We also have fractional numbers. For example, the decimal number 639.234 represents the polynomial $6 \times 10^2 + 3 \times 10^1 + 9 \times 10^0 + 2 \times 10^{-1} + 3$

$\times 10^{-2} + 4 \times 10^{-3}$. So is the case with any positional number system of the integer radix $r \geq 2$.

4.2.8 Base 2 and mode of computation/manipulation

It is interesting to note that while all hardware representation of any number or information involving numerical, alphabetical (nonnunerical), and special characters is in binary, the mode of computation/manipulation/arithmetic could be in a base/radix 2 or different from 2. We have binary[7], binary-coded decimal (BCD)[8], extended binary-coded decimal interchange code (EB-CDIC), octal[9], hexadecimal[10], variable radix, negative radix number systems in which computation/arithmetic is/could be done (Alam and Sen 1996; Krishnamurthy 1965, 1970a, 1970b, 1970c, 1970d, 1971a, 1971b; Krishnamurthy and Nandi 1967, Nandi and Krishnamurthy 1967; Metropolis and Ashenhurst 1963; Sankar et al. 1973a, 1973b). *Underlying this mode is always the binary system and nothing else.* The implementation of the foregoing arithmetic in a digital computer could be purely through software programs or firmware[11]. In firmware, the hardwired instructions implement the mode of computation/arithmetic while in software, the programming instructions written by a user/application programmer or a system programmer taking the features of the computer into account implement the mode of computation/arithmetic. Whichever be the implementation, the hardware computer has every information in a binary (i.e., base 2) form and not in any other base. Out of all the possible radix systems, it is the base 2 system in which any number can be written in the form $\Sigma d_n 2^n$, that stands not only the tallest but also the only form understandable by the hardwired instructions of the digital computer.

---

[7] used in HEC 2M, a British computer during late 1950's and early 1960's
[8] used in some IBM computers of 1960s
[9] used in URAL, a Russian computer existing during early 1960's.
[10] used in DEC 1090, a Digital Equipment Corporation computer (American) existing during late 1970's and early 1980's.
[11] Software implemented in hardware; To change/modify firmware, one needs to modify the electronic circuitry; consequently a firmware cannot be easily modifed while a software program can be easily changed. The execution of a software program takes more time than the corresponding firmware program. However, too frequently occurring functions and arithmetic operations are often implemented in firmware that saves significant amount of computing time.

4.2.9 Conversion from one base to any other and error

Conversion of a number in one base β to that in another base σ may be accomplished easily (Alam and Sen 1996). The number of symbols in base β may be greater than or less than those in base σ. If β = 10 (decimal) and σ = 16 (hexadecimal), then to convert the decimal number $(428.31)_{10}$ to the corresponding hexadecimal number, we may use *repeated division* on the integral part 428 and *repeated multiplication* on the fractional part .31 as follows.

428/16 = 26 remainder C; 26/16 = 1 remainder A; 1/16 = 0 remainder 1; .31 × 16 = 4.96, integer part = 4; .96 × 16 = 15.36, integer part = 15 = F; .36 × 16 = 5.76, integer part = 5; . . . Hence the corresponding hexadecimal number is 1AC.4F5 . The decimal equivalent (in Matlab form) of this hexadecimal number up to 3 hexadecimal places is

d = 1*16^2 +10*16^1+12*16^0 + 4*16^-1+15*16^-2+5*16^-3
= 428.3098.

Observe that we may not always have exact conversion from one base to another base. This nonexactness would introduce conversion error. There is yet another method called the *polynomial evaluation method* for conversion from one base to another. The foregoing decimal number 428.31 can be written, in this method, as $4 \times 10^2 + 2 \times 10^1 + 8 \times 10^0 + 3 \times 10^{-1} + 1 \times 10^{-2} = 4 \times A^2 + 2 \times A^1 + 8 \times A^0 + 3 \times A^{-1} + 1 \times A^{-2}$ (in hexadecimal number system) = $4 \times 64 + 2 \times A + 8 + 3 \times A^{-1} + 1 \times A^{-2}$ (in hexadecimal) ≈190 + 14 + 8 + .4CCCB + .028 (in hexadecimal) ≈ 1AC.4F5 . . . To convert a binary number, we may use the polynomial evaluation method. To convert the binary number $(1101.101)_2$ into decimal, we may write the polynomial $1 \times 2^3 + 1 \times 2^2 + 0 \times 2^1 + 1 \times 2^0 + 1 \times 2^{-1} + 0 \times 2^{-2} + 1 \times 2^{-3}$ (in decimal) = 8 + 4 + 0 + 1 + .5 + 0 + .125 (in decimal) = 13. 625 (in decimal).

Arithmetic in various bases (other than decimal) is exactly similar to decimal arithmetic. When working in base β, we carry and borrow β's (rather than 10's). The foregoing examples do illustrate this aspect.

4.2.10 Complementation

Addition of two nonnegative numbers is no problem while subtraction of two nonnegative numbers could be involved and costly in terms of circuit design and actual hardware. Most computers, both general and special purpose, perform subtraction by adding the complement of the number being subtracted. Thus, borrow and other circuits connected with subtraction are eliminated and cost is reduced. Consider the decimal subtraction 23 – 17.

Here −17 is represented as 100 − 17 = 83 which is then added to 23. That is 23 + 83 = 106. The extra digit 1 is discarded to see that adding the complement (called the true or, equivalently, the radix complement of 17) of 17 to 23 is exactly the same as subtracting 17 from 23 by conventional method.

Another way is to represent the decimal number −17 as (100 − 1) − 17 = 82 (called the digit, or, equivalently the radix − 1 complement of 17) which is then added to 23 and the resulting extra digit is brought around and added to the sum to get the correct result. Thus we have 23 + 82 = 105 and then 05 + 1 = 06 which is the correct result.

If we subtract a bigger number from a smaller number, we would get the result which is negative in a complement (digit or true depending on the specific computer implementation) form. The foregoing two identical procedures may be implemented in binary or, as a matter of fact, in any other number system.

The true or the digit complement is not of much use with the decimal number system since the computations of these complements are equally difficult. For the binary number system, however, the digit or, equivalently, one's complement is obtained merely by reversal of 1's and 0's. For example, for the subtraction of the binary number 10001 from the binary number 10111, we compute 10111 + 01110 = 100101; the extra left-most (most significant) digit 1 is brought around and added to the right-most (least significant) digit to obtain the correct result, viz., 00110. This process simplifies both subtraction and division.

Most computers perform subtraction by complementing the subtrahend and adding it to the minuend. Thus the computer can add, subtract, multiply, and divide by the simple process of add and shift operations.

4.2.11 Computer word

The main (executable) memory of the computer can be thought of as having a large number of fixed-length locations called words each of which can store a sequence of bits 0 and 1. The word length varies from computer to computer in the range 8 to 64 bits in general. The IBM 360 as well as IBM 370 computer words were 32 bit long while the DEC1090 computer word was 36 bit long. All these systems are main frames and obsolete to-day and have possibly become museum pieces.

Binary point has no explicit representation The representation of a binary point in a computer memory location or in a computer CPU register is not done in any explicit way. The point is assumed in binary number representations, viz., the fixed-point representation and the floating-point representation. Most computers store numbers in both ways.

4.3 Fixed- and floating-point representation and a arithmetic

The fixed-point representation assumes that the radix point is always fixed in one position of the memory word. If we imagine the point fixed at the extreme left, then all numbers are positive or negative fractions. On the other hand, if we regard the binary point to be fixed at the extreme right of the word, then all the numbers are positive or negative integers. These number representations are fixed-point representations. The arithmetic operations employed with these representations are termed as fixed-point arithmetic. Most computers currently being manufactured represent binary integers in fixed-point form.

Consider the binary integers, i.e., the binary numbers in which the binary point is imagined to be present at the right-most end of the computer word. The sign of a binary integer can be treated in any of the three forms — sign-and-magnitude, 2's complement, and 1's (digit) complement. The left-most bit of the computer word is usually used to represent the sign of the binary number.

4.3.1 Sign-and-magnitude form

The left-most bit is the sign bit. If the left-most bit is 0, then the number is positive. If it is 1 then the number is negative. The ordered sequence of bits following the sign bit represents the magnitude of the binary integer. In 32-bit computer words, the range of numbers representable is $[-(2^{32-1} - 1), (2^{32-1} - 1)] = [-2147483647, 2147483647]$. Zero is represented as **00000000** or 1 **0000000** 000, where **0** (bold) represents 0000. In a 32 bit computer, if two words are used to represent a fixed-point number then the range will be $[-(2^{64-1} - 1), (2^{64-1} - 1)]$.

4.3.2 2's and 1's complement forms

The nonnegative integers $\leq 2^{32-1} - 1$ are represented exactly in the same way as in the sign-and-magnitude form. The representation for the largest non-negative integer is 0 **1111111** 111, where **1** (bold) represents 1111. The negative integers in $[-2^{32-1}, -1]$ are represented by a 1 in the left-most bit (sign bit) and the 2's complement of the binary magnitude in the 32 – 1 bits following the sign bit. In 2's complement, unlike the sign-and-magnitude representation, 0 (zero) has only one representation, viz., all 32 bits 0. In fact, all numbers in $[-2^{32-1}, (2^{32-1}-1)]$ have a unique representation. Observe the lack of symmetry, i.e., -2^{32-1} is representable but 2^{32-1} is not.

Consider, for example, 6 bit word. This word length, in 2's complement, implies that the magnitude of the result of add/subtract operation should be ≤ $2^5 = 31$. Otherwise an actual overflow would occur and the result that remains in the 6 bit word would be wrong. The addition/subtraction of numbers in 2's complement notation is illustrated as follows.

+6	000110		−6	111010
−3	111101		+3	000011
+3	1 000011		−3	111101 ← Both the results are correct in 2's complement form
	↖ (overflow bit) discard			

The simplicity of the rules for add/subtract operations in the 2's complement notation as well as easy hardware implementability have made 2's complement notation a preferred one in many computers. For details on add/subtract operation in 1's complement notation as well as multiply/divide operartions in 2's complement and 1's complement notations, refer Alam and Sen (1996) and Rajaraman and Radhakrishnan (1983).

4.3.3 Floating-point representation of numbers and arithmetic

The floating-point representation of numbers corresponds closely to "scientific notation"; each number is represented as the product of a number with a radix point and an integral power of the radix. One bit of the word is for the sign of mantissa, e bits of the word for the exponent while f bits for the mantissa or, equivalently, significand (Forsythe and Moler 1967) as in Figure 4.1.

S	**E**	**F**
1 bit	**e bits**	**f bits**
Sign	**Exponent**	**Fraction (= Mantissa or Significand)**

Figure 4.1: Floating-point number format

The exponent bits (usually in excess 2^{e-1} code) represent the actual integer E The fraction (mantissa) bits represent the fraction F, where $0 < F < 1$. The number in the computer word would be $\pm F \times 2^E$. In other schemes, the value is taken to be $\pm F \times B^E$ for some constant B other than 2. IBM 360/370 computers use B = 16. Here we will consider B = 2. The exponent may be positive or negative. The sign bit represents the sign of the mantissa. The exponent expressed in excess 2^{e-1} code takes care of the sign of the exponent. If all the e bits are 0 then these bits will represent the actual exponent $-2^{e-1} = -128$ when the number of bits e = 8, i.e., the actual multiplier is $2^{-128} \approx 0.350 \times 10^{-45}$. If the leftmost bit (most significant) bit of e bits is 1 and the rest are zero then these bits will represent the true exponent $2^{e-1} - 128 = 0$ when the number of bits e = 8.

4.3.3.1 Dwarf and machine epsilon

In a 32 bit word, if one bit is for sign, 8 bits for exponent, and 23 bits for mantissa then the concept of the dwarf and the machine epsilon (Sen 2003) is important. The smallest representable number which is just greater than 0 in magnitude is called the dwarf (Figure 4.2). It would be, allowing **0** (bold) = 0000,

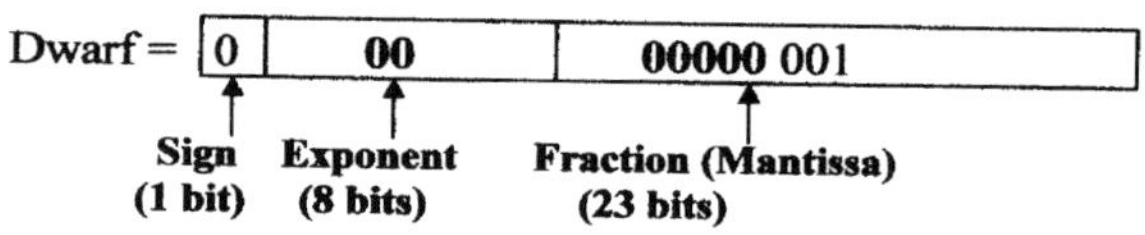

The true exponent is 2^{-128} and fraction is 2^{-23}. Hence the value of the dwarf is $2^{-151} \approx 3.503246160812043 \times 10^{-46}$. (In excess-128 code, 00 will represent $2^{0-128} = 2^{-128}$ as the actual exponent.)

Figure 4.2: The dwarf (smallest representable floating-point number just greater than 0) in a 32-bit word with 23 bit mantissa and 8 bit exponent

The machine epsilon (Figure 4.3) is the smallest number that a computer recognizes as being very much bigger than zero (as well as the dwarf) in magnitude. This number varies from one computer to another. Any number below the machine epsilon, when added or subtracted to another number, will not change the second number. It is represented, allowing **0** (bold) = 0000 (a block of four bits), as

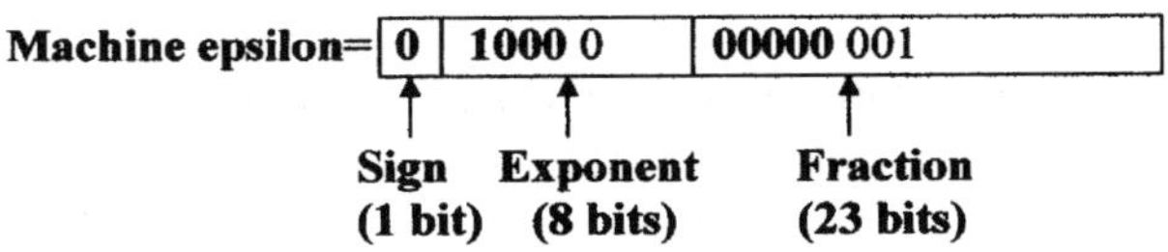

True exponent = 2^0 (in excess-128 code), machine epsilon is $2^{-23} = 1.19\times10^{-7}$. (In excess-128 code the exponent is in $[2^{-128}$ to $2^{127}]$).

Figure 4.3: Machine epsilon (the smallest number recognized by the computer as very much greater than zero (as well as dwarf) in magnitude and when added to 1 produces a different number)

During computation if a numerical value α < machine epsilon (and, of course, larger than the dwarf) then adding this value α to another value b will keep the result b only. Such a situation may lead to an infinite loop if b is tested against another value c. A legal way for jumping or breaking out of the loop is shown in the following MATLAB program called *mcepsilon*:

```
%mcepsilon
eps=1; format long;
  for n=1:500
    eps=eps/2; if (1+eps) <=1, eps=eps*2; break; end
  end
  n, eps
```

The machine epsilon (in double precision) is eps = 2.220446049250313e–016. It is obtained when the number of terms is n =53. This Matlab program shows a way of estimating the machine epsilon of a specified machine (here in double precision), in which the *for* loop is executed for sufficiently large number of times.

Using the Matlab hexadecimal format, viz., format hex instead of the decimal format, viz., format long, we would be able to see that the machine epsilon when added to any value in between ≥ 1 and < 2 will change the value. But it when added to any value ≥ 2 will not change the value. For **example, after running the foregoing mcepsilon (Matlab) program, we obtain**

n = 53, eps = 2.220446049250313e-016.

Now using the following Matlab commands

```
» format hex
```

» **eps**

we obtain **eps = 3cb0000000000000.** The following Matlab command

» **1+eps**

gives the result **ans = 3ff0000000000001** whereas the representation of 1 in hexadecimal format is **3ff0000000000000** which is different in the last hexadecimal digit. The command

» 1.9999999999999+eps

produces **ans = 3ffffffffffffe3f**. The command

» **1.9999999999999**

gives **ans = 3ffffffffffffe3e** which differs in the last (least significant) hexadecimal digit. The commands

» **2+eps**

and

» **2**

produce the same result **ans = 4000000000000000**. Also, the commands

» **3+eps**

and

» **3**

produce the same result **ans =4008000000000000**

The Matlab program called dwarf for determining the dwarf may be written as follows.

```
%dwarf
eps=1; format long;
  for n=1:1500
    dwarf=eps;
    eps=eps/2;
```

```
  if eps==0, break; end;
end
n, dwarf
```

The value of dwarf (in double precision) is given as **dwarf = 4.940656458412465e–324** and the corresponding number of terms **n = 1075**.

The floating-point representation provides much larger range of values to be represented unlike the fixed-point representation. The disadvantage of the floating-point notation is that we do not obtain as many as k – 1 significant bits in one word for a k-bit word computer.

4.3.4 Normalized form and limitation

A condition is often imposed to avoid ambiguity/nonuniqueness in the floating-point representation of numbers. The condition is that the most significant digit of the mantissa is always nonzero. Such a floating-point number is in the *normalized form*. Unfortunately, normalization does not permit zero to be represented. A natural way to represent zero in a 32-bit (single precision) machine with 8 bit exponent is with $1.0 \times 2^{0-128}$ since this preserves the fact that the numerical ordering of the nonnegative real numbers corresponds to the lexicographical ordering of their floating point representations. This ordering is based on the assumption of the conventional arrangement where the exponent is physically stored to the left of the mantissa (fraction). In an 8 bit field, this implies that only $2^8 - 1 = 255$ values are available for use as exponent since one is reserved to represent 0. For further details and for the floating-point arithmetic, refer (Goldberg 1991; Alam and Sen 1996).

4.3.5 Other representations

Floating-point representations have a radix β *which is always taken as even* and a precision p. There are several other representations, viz., *floating slash* and *signed logarithm* (Matula and Kornerup 1985; Swartzlander and Alexopoulos 1975). However, the floating-point representation is the most widely used format in almost all computers.

To represent a negative binary number, 2's complement or 1's (digit) complement is used. From electronic switching point of view, such a complementation is easy and fast (immediate).

4.3.6 Floating-point arithmetic and consequences

Addition (⊕) To add two normalized floating-point numbers of the same sign, the higher of the two exponents is chosen for the result, and the digits of the other mantissa (significand) are suitably shifted. The choice of the higher exponent is based on the theory of error analysis. If the addition results in a mantissa greater than 1, then the resulting floating-point number is shifted to the right by one digit and the exponent is increased by 1 if the exponent is within the range. Else, the result overflows. For example, consider the mantissa of length 2 + 1 digits and exponent of length 1 + 1 digits. If the floating-point numbers are a = (.94, 9) and b = (.17, 9), then a ⊕ b will overflow. The problem of adding two numbers of opposite signs may be treated as that of subtraction.

Subtraction (Θ) Here also the higher exponent is retained. The resulting floating-point number, when normalized, might result in underflow (Demmel 1984; Krishnamurthy and Sen 2001). Assuming the same length of mantissa and that of exponent as in addition, the result a Θ b, where the floating-point numbers a = (.53, –9) and b = (.51, –9), will underflow.

Multiplication (⊗) To multiply two normalized floating-point numbers, mantissas are multiplied and exponents are added and the resulting floating number is normalized, rounded, and the exponent appropriately adjusted. Here the result may overflow or underflow.

Division (∅) In dividing one normalized floating-point number by another, the mantissa of the dividend is divided by that of the divisor, the exponent of the divisor is subtracted from that of the dividend, the resulting mantissa is then normalized (to make the most significant digit nonzero) and rounded, and the exponent is appropriately adjusted. Here also, like multiplication, the result may underflow or overflow.

Consequences Floating-point arithmetic is performed almost entirely with normalized floating-point numbers. The resulting floating-point numbers are almost always in normalized floating-point forms. Since the arithmetic is erroneous (inexact), the computed result always contains noise. Consequently, the floating-point addition and multiplication are only commutative; both the associative and distributive laws do not hold. If a, b, and c are three normalized floating-point numbers, then

$$a \otimes (b \oplus c) \neq a \otimes b \oplus a \otimes c; \; a \oplus a \oplus a \neq 3.0 \otimes a.$$

A method of avoiding nonassociative analysis for floating-point operations is difficult and is yet to be sufficiently explored. A method known as *inverse error analysis* due to C. Lanczos and W. Givens has been extensively used by Wilkinson (1963, 1965). One is required, in this method, to determine how small a change in the data of a problem would be necessary to cause the computed answers to be the exact solution of the changed problem. Consider, for example, the quadratic equation $1.00000x^2 - 6.00000x + 8.99999 = 0$. If the computed roots are 3, 3, then we can check that these are the exact roots of the equation $0.9999997x^2 - 6.0000001x + 9.0000030 = 0$. Since the coefficients in the equation differ from those in the former by not more than 1 ulp (unit in the last decimal place – defined in Sections 4.9.15 and 4.9.16), the aforesaid roots may be considered fairly good for the former equation. The other method known as *direct error approach* asks how wrong the answer is as the solution of the problem with their given data. The inverse error approach can permit, unlike the direct error approach, us to easily continue to use associative operations in many large matrix or polynomial problems.

4.3.7 Magnitude relation between floating-point numbers

The equality of two floating point numbers cannot be easily established. In iterative procedures with respect to infinite algorithms, we can only test $|x_{i+1} - x_i| \le \varepsilon$ (absolute error test), where x_i is the i-th iterate and ε is a suitably chosen (degree of approximation) positive real number. However, we more often or almost always use the test $|x_{i+1} - x_i| \le \varepsilon|x_{i+1}|$ (relative error test). This test will indicate whether x_{i+1} is approximately equal to x_i.

To compare the relative values of any two floating-point numbers A =(a, ea) and B = (b, eb) in radix β, the following definitions (Krishnamurthy and Sen 2001; Wijngaarden 1966; Knuth 1969) are useful. Let $\prec$, $\succ$, $\approx$, and $\cong$ denote "definitely less than", "definitely greater than", "essentially equal to", and "approximately equal to", respectively. Then the relations are

$$A \prec B \text{ iff } B - A > \varepsilon.\max(\beta^{ea}, \beta^{eb});\ A \cong B \text{ iff } |B - A| \le \varepsilon.\max(\beta^{ea}, \beta^{eb});$$

$$A \succ B \text{ iff } A - B > \varepsilon.\max(\beta^{ea}, \beta^{eb});\ A \approx B \text{ iff } |A - B| \le \varepsilon.\min(\beta^{ea}, \beta^{eb});$$

Observe that $\approx$ is stronger than $\cong$. Consider, for example, A =(.401, 1), B = (.404, 1), ε = .001, β = 10. Then $A \prec B$ since B – A = .03 > .01. If now B = (.402, 1), then $A \cong B$ and $A \approx B$ since $|B - A|$ =.01 ≤ .01. Allowing ε =.0001, we have $A \prec B$ but the relations $A \cong B$ and $A \approx B$ do not hold. Thus the zero in floating-point numbers depends on the choice of ε. Hence it is not possible to define an exact zero. Consequently, the following relations can be proved.

$$|A - B| \le \varepsilon|A| \text{ and } |A - B| \le \varepsilon|B| \Rightarrow A \approx B,$$
$$|A - B| \le \varepsilon|A| \text{ or } |A - B| \le \varepsilon|B| \Rightarrow A \cong B.$$

For normalized floating-point numbers A and B with $\varepsilon < 1$, the following relations hold.

$$A \approx B \Rightarrow |A - B| \le \beta\varepsilon|A| \text{ and } |A - B| \le \beta\varepsilon|B|,$$

$$A \cong B \Rightarrow |A - B| \leq \beta\varepsilon|A| \text{ or } |A - B| \leq \beta\varepsilon|B|,$$
$$A \prec B \Rightarrow B \succ A \text{ as well as } A < B;\ A \approx B \Rightarrow A \cong B.$$

4.3.8 Unnormalized floating-point and significance arithmetic

Unnormalized floating-point arithmetic Normalizing all floating-point numbers will not be always favourable to attempt the maximum possible accuracy for a given precision. Sometimes it may tend to imply that the computations are more accurate than they really are. If, for example, A Θ B is normalized, when A = (.514267, 1) and B = (.514146, 1), then A Θ B = (.121000, −2); the information about the possibly greater inaccuracy of the result is suppressed; if the result were (.000121, 1), the information would not be suppressed. Ashenhurst and Metropolis (1959, 1965) as well as Metropolis and Ashenhurst (1963) suggested the unnormalized arithmetic to retain the information.

The rules for unnormalized arithmetic are as follows. Let za be the number of leading zeros in the fractional mantissa (significand) a of the floating-point number A = (a, ea) while zb be the number of leading zeros in the fractional mantissa (significand) b of the floating-point number B = (b, eb). Also, let p be the precision so that za is the largest integer ≤ p with $|a| < \beta^{-za}$, where β is the radix. Then addition and subtraction are carried out as in the normalized floating-point arithmetic except that here normalization is suppressed while multiplication and division are performed in the same manner except that the result is scaled left or right so that max (za, zb) zeros appear. For unnormalized arithmetic, the rules (Krishnamurthy and Sen 2001) for determining the exponent are as follows.

$$e_{A \oplus B,\, A \ominus B} = \max(ea, eb) + (0 \text{ or } 1),$$
$$e_{A \otimes B} = ea + eb - \min(za, zb) - (0 \text{ or } 1),$$
$$e_{A \oslash B} = ea - eb - za + zb + \max(za, zb) + (0 \text{ or } 1).$$

An unnormalized zero will be produced when the result of computation is zero. The relations ≺, ≻, ≈, ≅ hold also for unnormalized numbers. Although there is no clear guidelines based on theoretical analysis to choose from among the normalized or unnormalized systems of arithmetic, the IEEE arithmetic/standard has been the most widely implemented arithmetic on computers.

Significance arithmetic Besides interval arithmetic (Section 4.9.14), another approach is to use significance arithmetic in which, like the interval arithmetic, a pair of numbers is used to represent the center and the half-length of the interval containing the quantity (Goldstein 1963; Dwyer 1951).

Other arithmetic There are problems where we may like to use rational or integer arithmetic, or p-adic or multiple modulus residue arithmetic for error-free/high-accuracy computation (Crandall and Fagin 1994; Matula and Kornerup 1985; Gregory and Krishnamurthy 1984; Lakshmikantham et al. 1997; Sen and Jayram 1980; Rao 1975; Venkaiah 1987; Venkaiah and Sen 1987, 1988, 1990).

4.3.9 Operations in multiple precisions

When a number is stored in one word (e.g., 32 bit word) of the memory of a computer, the number is called a single-precision number. When the single-precision arithmetic is not enough to get a desired accuracy, the precision can be increased using two (or more) words of the memory to represent each number. In such a case appropriate algorithms/subroutines have to be designed to do the arithmetic. This is known as double- (or multiple-) precision arithmetic. Both the fixed-point numbers as well as the floating-point numbers can be in single-, double-, multiple-, and variable precisions. A multiple-precision operation (add, subtract, multiply, or divide) takes several times more time than the corresponding single-precision operation.

For the *addition* of multiple-precision numbers, each of the operands (numbers) can be segmented to the standard word size (say, 32 bits). The individual segments can be added together with the carry from the previous segments (Krishnamurthy and Sen 2001). For example,

$$(\Sigma a_i) + (\Sigma b_i) = \Sigma(a_i + \Sigma b_i)$$

where the summation is over i = 1(1)n. The *subtraction* also is similarly carried out.

For the multiple-precision *multiplication*, on the other hand, cross-products have to be computed and these have to be added using multiple-precision addition. For instance,

$$(\Sigma a_i)(\Sigma b_i) = a_1b_1 + a_1b_2 + \ldots + a_1b_n + a_2b_1 + a_2b_2 + \ldots + a_2b_n + \ldots + a_nb_1 + a_nb_2 + a_nb_n,$$

where the summation is over i = 1(1)n.

For the multiple precision *division*, we assume that a facility for dividing a double-precision segment by a single-precision segment is available. The problem is involved since, by segmenting, $(\Sigma a_i)/(\Sigma b_i)$ cannot be expressed exactly as a sum of the individual ratio of segments. Very efficient *divide-and-correct algorithms* have been suggested for this purpose (Krishnamurthy 1965; Krishnamurthy and Nandi 1967; Krishnamurthy and Sen 2001; Knuth 1969). These algorithms arrive at a trial quotient by dividing a double-precision segment of the dividend by a single-precision appropriately rounded segment of the divisor. The quotient is then corrected by ± 1 according to certain rules based on the sign of the round-off of the divisor and the sign of the quotient.

One may use the *binomial expansion* to form

$$a/(b + \varepsilon) = (a/b)(1 - \varepsilon/b + \varepsilon^2/b^2 - \varepsilon^3/b^3 + \ldots),$$

where b is an appropriately chosen single-precision segment and ε is a small number compared to b. This approach is more expensive than the divide-and-correct procedures.

One may also use *fast functional iterative schemes* for the division (Krishnamurthy 1970a-d, Krishnamurthy and Sen 2001). Here we compute a/b without remainder, where a and b are p-digit fractions in normalized form, i.e., $(1/\beta) \le a, b \le 1$ (β is the radix). We then construct a sequence of multipliers m_i $i = 0(1)n$ such that $b\Pi m_i$, where $i = 0(1)m$ converges to a definite limit c for some reasonably small n. The dividend a is also simultaneously multiplied by the same sequence of multipliers m_i. Allowing $a = y_0$ and $b = x_0$, the iterative procedure

$$x_{i+1} = m_i x_i,\ y_{i+1} = m_i y_i,\ i = 0, 1, 2, \ldots, \text{ till } |y_{i+1} - y_i|/|y_{i+1}| \le \varepsilon \quad (\varepsilon = .5 \times 10^{-15}, \text{ say}) \text{ such that } x_i = c,\ y_i = cq,\ q = y_i c^{-1}.$$

The procedure thus needs a selection of m_i and multiplications, and a final step to multiply by c^{-1}.The m_i are selected to be easily computable and, at the same time, c^{-1} is a convenient integer such that $c^{-1}y$ is easily computed. Krishnamurthy (1970a) has shown that the best choice is

$$m_i = (2c - x_i)/c,\ \ 0 < x_0 < 2c.$$

Consider, for example, a/b = 1/.8. Here $x_0 = .8$, $y_0 = 1$ $c = 1$, $m_0 = 1.2$. Accuracy desired is $\varepsilon =.5 \times 10^{-15}$. Hence $x_1 = m_0x_0 = 1.2 \times .8 = .96$, $y_1 = m_0y_0 = 1.2 \times 1 = 1.2$, $m_1 = (2c - x_1)/c = 1.04$. Since the relative error $|y_1 - y_0)|/|y_1| = .2/1.2 = .1667 > \varepsilon$, we go to the second step. $x_2 = m_1x_1 = 1.04 \times .96 = .9984$, $y_2 = m_1y_1 = 1.04 \times 1.2 = 1.2480$, $m_2 = 1.0016$. Since $|y_2 - y_1)|/|y_2| > \varepsilon$, we proceed to the third step. $x_3 = m_2x_2 = 1.0016 \times .9984 = 1$, $y_3 = m_2y_2 = 1.0016 \times 1.2480 = 1.25$, $m_3 = 1$. Since $|y_3 - y_2)|/|y_3| > \varepsilon$, we go to the fourth step. $x_4 = m_3x_3 = 1 \times 1 = 1$, $y_4 = m_3y_3 = 1 \times 1.25 = 1.25(2c - x_1)/c = 1.25$, $m_4 = 1$. Since $|y_4 - y_3)|/|y_4| \le \varepsilon$, we stop the iteration. Thus $q = c^{-1}y_3 = y_3 =1.25$ is the required answer.

4.3.10 Important points to remember in floating-point computation

In floating-point computations, the important points that one should remember are as follows.

(i) Try to select those algorithms which involve least number of arithmetic operations (i.e., the least computational complexity) as these would result in least error, in general.

(ii) Use multiple precision whenever needed but not without sufficient reason, as this is expensive (computational complexity increases).

(iii) Mathematically identical problems may be numerically quite different.

(iv) Whenever a subtraction is encountered involving two nearby quantities, exercise sufficient care. If possible, reformulate the problem/subproblem. Otherwise do the subtraction before performing the other operations.

4.3.11 Significance of a quantity/number

The quantity $\sigma(Q) = \log_\beta(1/\text{relative error}) = \log_\beta(|Q|/|Q - Q'|)$ is defined as the *significance* of the quantity/number Q. The lower integral part of $\sigma(Q)$, i.e., $\lfloor\sigma(Q)\rfloor = \lfloor\log_\beta(|Q|/|Q - Q'|)\rfloor$ gives the number of significant digits up to which the quantity/result Q is correct. The quantity $\lfloor\log_\beta(1/\text{absolute error})\rfloor = \lfloor\log_\beta(1/|Q - Q'|)\rfloor$, on the other hand, gives the number of decimal digits up to which the quantity/result is correct.

Consider the quantity Q of higher order accuracy as 253.2315, the quantity Q′ of lower order accuracy as 253.1891, and the base $\beta = 10$. Then the absolute error is $|Q - Q'| = 0.0424$, the relative error is $|Q - Q'|/|Q| = 1.6744 \times 10^{-4}$, the percentage error in Q is 0.0167, the number of significant digits up to which the quantity Q is correct is $\lfloor\log_{10}(1/\text{relative error})\rfloor = \lfloor 3.7762\rfloor = 3$, the number of decimal digits up to which the result is correct is $\lfloor\log_{10}(1/\text{absolute error})\rfloor = \lfloor 1.3726\rfloor = 1$.

If $Q=0.0003781$, $Q'=0.0002989$, $\beta=10$ then the absolute error in Q is 7.9200×10^{-5}, the relative error in Q is 0.2095, the percentage error in Q is 20.95, the significance of Q is 0.6789, the number of decimal digits as well as that of significant digits up to which Q is correct is $\lfloor 4.1013\rfloor=4$ and $\lfloor 0.6789\rfloor=0$, respectively. The former percentage error, viz, 0.0167 is much less than the later one, viz., 20.95. This implies that the earlier result Q′ is much more accurate.

4.3.12 Error in arithmetic operations

Let Q_1 and Q_2 be two given approximate quantities. Both have a certain order (usually the same order) of error associated with them. The relative error in *addition,* i.e., in $Q = Q_1 + Q_2$, will be the order of the larger relative error when adding approximate quantities.

The relative error in *subtraction* will be greater than each of the two relative errors. If Q_1 and Q_2 are nearly equal then the relative error in Q, i.e., $\Delta Q/|Q|$ will be large and consequently a large number of significant digits representing Q will be lost. Hence whenever possible, try to avoid subtracting two nearly equal (nearby) numbers.

The relative errors are added when *multiplying/dividing* two approximate quantities. For further details, refer Krishnamurthy and Sen (2001), Alam and Sen (1996).

4.3.12.1 Is true error in computation non-decreasing

It may be noted that the error (implying error-bounds) in real quantities, like entropy (defined, in thermodynamics, as $\Delta Q/T$ where ΔQ is the change in heat energy and T is the absolute temperature) in physics can never be reduced by any operation – arithmetic or otherwise. Thus the error is monotonically nondecreasing (increasing, in general) under any arithmetic operation and this needs to be computed to ascertain the quality of the result.

However, the true error (never known) could be on the positive side or on the negative side of the exact quantity (never known). Hence, the true cumulative error (also never known) in the computation could be decreasing, i.e., less than each of the true errors in the quantities (inputs) involved in computation. The true cumulative error even could be zero.

For example, let the exact quantities happen to be 2.34567, 3.45678, and 4.56789 while the corresponding approximate quantities are 2.34563, 3.45680, and 4.56791. Then the sum of these three approximate quantities, viz., 10.37034 is errorfree.

4.3.12.2 Example of error in arithmetic operations

To illustrate the error in four arithmetic operations, consider $Q_1 = 12.3481$ with the error $\Delta Q_1 = 0.5 \times 10^{-3}$ and $Q_2 = 15.6238$ with the error $\Delta Q_2 = 0.5 \times 10^{-3}$. The two foregoing errors are absolute errors and have the same order.

In the physical world, unless human errors creep in, the same measuring instrument will always have the same order of error. Further, there exists usually no measuring instrument that gives an accuracy more than 0.005% (i.e., roughly 4 significant figures/digits).

The relative error in Q_1 is $\Delta Q_1/Q_1 = 0.4049 \times 10^{-4}$, that in $Q_2 = \Delta Q_2/Q_2 = 0.3200 \times 10^{-4}$, the absolute error in $Q = Q_1 + Q_2$ is $\Delta Q = \Delta Q_1 + \Delta Q_2 = 0.1 \times 10^{-2}$, the relative error in Q is 0.3575×10^{-4}. The absolute error in $Q = Q_1 - Q_2$ is 0.1×10^{-2} (observe that the error has been added and not subtracted), the relative error in $Q = Q_1 - Q_2$ is 0.3053×10^{-3} (observe that the relative error in the subtraction is more than that in addition).

Subtraction of nearby numbers If the two quantities Q_1 and Q_2 are nearby then the relative error in $Q = Q_1 - Q_2$ will increase significantly, i.e., a large number of significant digits will be lost. Hence enough care should be taken while subtracting two nearly equal numbers. Attempt should be made to avoid subtracting two nearby numbers or higher precision (double or multiple precision) should be used to compensate the loss of significant digits. Instead of computing $Q_1 - Q_2$, we may compute $(Q_1^2 - Q_2^2)/(Q_1 + Q_2)$ for better accuracy if Q_1 is nearly equal to Q_2. But this type of replacement has other drawbacks in terms of computing and programming expenses. The relative error in $Q = Q_1Q_2$ is 0.7250×10^{-4} and that in $Q = Q_1/Q_2$ is 0.7250×10^{-4}.

The relative error, in multiplication, in $Q = Q_1Q_2$ is $\Delta Q/Q = \Delta Q_1/Q_1 + \Delta Q_2/Q_2 = 0.4049 \times 10^{-4} + 0.3200 \times 10^{-4} = 0.7249 \times 10^{-4}$. Thus the relative errors are added while multiplying. Hence the result cannot have more significant digits than the number of significant digits in the less accurate factor (quantity).

Since error implying error-bounds is always considered nonnegative and since any arithmetic operation is also always considered to produce error greater than or equal to the larger error in a factor, the relative errors are added in division $Q = Q_1/Q_2$ too just as in multiplication.

4.4 Error in function with approximate arguments (direct problem)

Let

$$f = f(x_1, x_2, \ldots, x_n).$$

Then

$$\Delta f = (\partial f/\partial x_1)\Delta x_1 + (\partial f/\partial x_2)\Delta x_2 + \ldots + (\partial f/\partial x_n)\Delta x_n.$$

Hence

$$\Delta f/f = (1/f)\Sigma(\partial f/\partial x_i)\Delta x_i,$$

where i varies from 1 to n. Given the errors Δx_i in the argument x_i, we can thus compute the absolute error Δf in the function f as well as the relative error $\Delta f/f$ in f. If $f(x_1, x_2) = x_1^3/x_2^5$, $\Delta x_1 = \Delta x_2 = 0.5 \times 10^{-4}$ then the absolute error in f is $\Delta f = (\partial f/\partial x_1)\Delta x_1 + (\partial f/\partial x_2)\Delta x_2 = (3x_1^2/x_2^5)\times 0.5 \times 10^{-4} + (-5x_1^2/x_2^6) \times 0.5 \times 10^{-4}$. If $x_1 = 7$ and $x_2 = 8$ then the absolute error in f is $\Delta f = 0.1028 \times 10^{-6}$ while the relative error in f is $\Delta f/f = 0.9821 \times 10^{-5}$.

Observe that errors implying error-bounds are conventionally represented as a nonnegative quantity. It can be shown that the relative error in the common (base 10) logarithm is about twice the absolute error in it. Further it can also be shown that the error in a logarithm may cause unacceptable (large) error in the corresponding antilogarithm (i.e., the number).

Consider the *physical problem*: What are the errors – absolute as well as relative – in the power (= p) dissipated in a (r =) 10 Ohm resistor that carries a current of (I =) 3 A? The resistance-measuring instrument (ohmmeter) used can measure resistance up to 100 Ohms while the electric current-measuring instrument (ammeter) used can measure current up to 10 A. Both the instruments have the accuracy 0.1%. (This accuracy implies that the absolute error in ohmmeter is 0.1 Ohm and that in ammeter is 0.01 A). The absolute error in power $p = i^2r = 3^2\times 10$ Watt=90 Watt can be given as $\Delta p=(\partial p/\partial i)\Delta i + (\partial p/\partial r)\Delta r = 2ir\Delta i+i^2\Delta r=2\times 3\times 10\times 0.01+3^2\times 0.1=1.5$ Watt.

This absolute error implies that the exact power dissipated lies in [88.5 Watt, 91.5 Watt] and this exact value is never known and will never be known. The relative error is $\Delta p/p=1.5/(i^2r)=1.5/(3^2\times 10)=1.67\%$.

4.5 Error in arguments with prescribed accuracy in function (inverse problem)

The problem of obtaining the allowable errors in the arguments $x_1, x_2, \ldots, x_n$ when the error Δf in the function f is specified is indeterminate since there is only one equation for Δf and there are n unknowns $\Delta x_1, \Delta x_2, \ldots, \Delta x_n$. So, we use the *principle of equal effects* which is, in the real-world situation, quite reasonable. For example, associated with each measuring instrument there is an order of error which is fixed. When this instrument is used to measure a quantity several times or different similar quantities once each (or more than once each) then the order of error in each of the measurements will remain the same. The principle of equal effects is thus justified. It assumes that the values $(\partial f/\partial x_i)\Delta x_i$, $i=1(1)n$ are all equal. Hence

$$\Delta f=n(\partial f/\partial x_i)\Delta x_i \text{ or } \Delta x_i=\Delta f/[n(\partial f/\partial x_i)],\quad i=1(1)n.$$

Let the value of the function $f(x_1, x_2) = x_1^2 \sin x_2$ be required to three decimal places (Sen 2003; Krishnamurthy and Sen 2001). We find the permissible errors in x_1 and x_2 when x_1 is approximately 10 and x_2 is approximately 25 as follows. Here $\Delta f = 0.5 \times 10^{-3}$, $x_1 = 10$, $x_2 = 25$, $n = 2$, $\partial f/\partial x_1 = 2x_1 \sin x_2 = -2.6470$, $\partial f/\partial x_2 = x_1^2 \cos x_2 = 99.1203$. Hence the permissible error in x_1 is $\Delta x_1 = \Delta f/[n(\partial f/\partial x_1)] = 0.9445 \times 10^{-4}$ (omitting the negative sign), and that in x_2 is $\Delta x_2 = \Delta f/[n(\partial f/\partial x_2)] = 0.2522 \times 10^{-5}$. As a particular case, for a function f(x) of one argument (one independent variable) x, the permissible error in x is $\Delta x = \Delta f/(df/dx)$. Thus, if $f(x) = 2\log_e x$ then $\Delta x = x\Delta f/2$. If $f(x) = e^{-x}$ then $\Delta x = e^x \Delta f$ (omitting the negative sign).

Consider the *physical problem*: The absolute error in power dissipated in 10 Ohm resistor carrying a current of 3A should not be more than 1 Watt. What are the allowable absolute errors in measuring current and resistance? Here Δp=1 Watt, i=3 A, r=10 Ohm, n=2, $\partial p/\partial i = 2ir = 2\times3\times10 = 60$, $\partial p/\partial r = i^2 = 3^2 = 9$. Hence the allowable absolute error in measuring the current is $\Delta i = \Delta p/[n\partial p/\partial i] = 1/[2\times60] = 1/120 = 0.0083$ A while that in measuring resistance is $\Delta r = \Delta p/[n\partial p/\partial r] = 1/[2\times9] = 1/18 = 0.0556$ Ohm.

4.6 Significance of a function

As we have already seen, the significance of the quantity Q is $\sigma(Q) = \log_{10}(1/\text{relative error in } Q) = \log_\beta(Q/\Delta Q)$ if the base of the number representation is β and ΔQ is the absolute error in Q. Thus, the significance of x is $\sigma(x) = \log_\beta(x/\Delta x)$. The significance of the function is $\sigma(f) = \log_\beta(f(x)/[(df/dx)\Delta x]$. If $f(x) = 2x^{0.5}$ and the base $\beta = 10$ then $\sigma(f) = \log_{10}(2x/\Delta x)$. If x is approximately 1 and $\Delta x = 0.5 \times 10^{-3}$ then the significance of the function (Sen 2003; Krishnamurthy and Sen 2001) is $\sigma(f) = 3 + \log_{10} 4 = 3.6021$ and the number of significant digits up to which the value of the function is correct is $\lfloor 3.6021 \rfloor = 3$.

Consider the *physical problem*: For a constant resistance r=10 Ohm, the power p dissipated across this resistance is a function of the electric current i which has the absolute error Δi=0.01 A. If i=3 A, the significance of the function p(i) is $\sigma(p) = \log_{10}(p(i)/[(\partial p/\partial i)\Delta i])$. Since i is approximately 3 A and $\Delta i = 0.01$ A, we have $\sigma(p) = \log_{10}(90/[60\times0.01]) = \log_{10}(90/0.6) = 2.1761$. Hence the number of significant digits up to which the numerical value of power is correct is 2 (the lower integral part of 2.1761).

4.7 Error in series approximation

A series is the sum of terms. A sequence, on the other hand, is the collection of terms. The sum of the terms in the sequence will be called a series. For example, $1, x, x^2/2!, x^3/3!, .., x^n/n!, ..,$ to ∞ is a (an infinite) sequence while $1 + x + x^2/2! + x^3/3! + .. + x^n/n! + ..$ to ∞ is a (an infinite or a power) series. The above sequence and the series are infinite. If there is a finite number of terms in a sequence (or in a series) then the sequence (or the series) is finite. The term 1 in the sequence (or the series) is the 0-th term and the term $x_n/n!$ is the n-th term of the sequence (or the series). One may, however, call 1 as the first term and $x^n/n!$ as the (n +1)st term.

The series computation involves the addition of terms. This *addition is not usually carried out by explicitly computing the value of each term and then adding them up.* It is carried out by expressing the (k+1)st term in terms of the k-th term and adding the (k+1)st term to the already computed sum up to k-th term. In the foregoing series, the n-th term is $t_n=x^n/n!$ and the (n+1)st term is $t_{n+1} = x^{n+1}/(n+1)!$. Hence the scheme for computing the value of the series

$$s = 1 + x + x^2/2! + x^3/3! + .. + x^n/n! + .. \text{ to } \infty$$

is

$$s_0 = t_0 = 1,\ x = \text{a given number (real or complex)},$$
$$t_{n+1}=t_n x/(n+1) \text{ and } s_{n+1}=s_n + t_{n+1} \quad n = 0, 1, 2, \ldots, \text{ till } |t_{n+1}|/|s_{n+1}| \le 0.5 \times 10^{-4}$$

The value of s_{n+1} after the execution of the foregoing scheme is the required value of the series correct up to 4 significant digits. If we desire the value of the series correct up to 4 decimal places then we replace, in the foregoing scheme, $|t_{n+1}|/|s_{n+1}|$ by $|t_{n+1}|$. Observe that $|t_{n+1}|/|s_{n+1}|$ is the relative error while $|t_{n+1}|$ is the absolute error introduced due to the truncation of the infinite series after (n + 1)st term.

Further, *in numerical computation, we should almost always compute the accuracy in terms of significant digits and not in terms of decimal digits.* In the foregoing computation we have assumed sufficiently large precision (word-length) of the computer used so that the rounding error is too small to affect the accuracy up to 4 significant digits.

For the purpose of a computer implementation, we omit the subscripts to save storage space and write the *computational scheme* as

$$s = t = 1,\ x = \text{a given number (real or complex)},$$
$$t=tx/(n+1) \text{ and } s=s + t \quad n = 0, 1, 2, \ldots, \text{ till } |t|/|s| \le 0.5 \times 10^{-4}$$

Here '=' is not mathematical equal to. '=' implies 'is replaced by'.

4.7.1 Speed of convergence

Some infinite series are fast convergent while others are slow convergent. Some diverges beyond certain range of values of the parameter x when the series is the function of x while some others are only conditionally convergent. The foregoing series is e^x and is fast convergent. To get an accuracy of 4 significant digits, if we do not need more than 5 or 6 terms of the infinite series for a specified range of values of x then the series is excellent. If, on the other hand, we need more than, say, 20 terms for a specified range of values of x then the series is not, in general, very desirable for numerical computation. The computation of $\log_e(1 + x)$ by the series $x - x^2/2 + x^3/3 - x^4/4 + x^5/5 - \ldots$, to $\propto$, ($|x| \le 1$ and $x \ne -1$) is clearly undesirable for values close to 1 since it takes too many terms and hence too much computation and consequent error.

4.8 Base 2 system: best in computer/communication

It is not difficult to observe that out of numerous possible number systems including those with a very large base, say 36 (needing 36 distinct symbols), the only number system that pervades whole of our electronic media (including the communication media) is the one with base 2. Not only the base 2 number system, but also the information representation is completely in binary. The most wonderful invention of the twentieth century, viz., the internet communication is most effectively performed in binary form with minimal (or no) error under noisy environment. In fact, for very noisy channels, two phases corresponding to two symbols is the best one could use.

Although the number 2 is even, it can be used as a base in the finite-field computation unlike other even numbers 4, 6, 8, . . . This is because 2 is the only prime which is even and the finite-field computation needs only primes (even or odd) as the bases. Observe that all other primes (infinity of them) are odd.

Before the advent of digital computers, i.e., before 1940's, $\log_e$ (natural logarithm), and $\log_{10}$ (common logarithm) were the ones most used and most dominant. During the digital computer age, $\log_2$ has gained at least the same importance as $\log_e$ and $\log_{10}$.

The whole of yes-no logic — the easiest one from comprehension point of view — represented by the base 2 symbols 0 and 1, pervades several other areas/problems such as the representation of a general tree as an equivalent binary tree, bisection for the solution of nonlinear equations, binary search, noisy channels, and common human psychology. One could think of several levels between yes and no and create multi-valued logic. Fuzzy set theory involves this kind of logic. However, the two-valued logic is the simplest of all. It can be seen that we cannot have one-valued logic. Thus, the enormous significance of 2 — more generally 2^n — has affected most walks of our lives. Just as the impact of the digital computer in the society is increasing rapidly, so is the impact of 2 or more generally 2^n among all numbers in various bases is growing. In fact, people in different disciplines have automatically made the decimal number system as an integral part of their lives. Many of them are now increasingly getting more familiar with base 2 number system as this system forms the core of all imaginable information including images in various colors in digital computers, in (computer) communication, and in embedded computing. All other numbers in various number systems fell behind and are of mainly academic interest.

4.9 IEEE 754 floating-point format

The arithmetic, say IEEE standard 754 binary floating-point arithmetic, that is often used with additional features, takes full advantage of binary representation in the hardware computer. The IEEE standard

(**http://www.cs.berkeley.edu/~wkahan/ieee754status/IEEE754.PDF**),

for example, specifies three formats — single, double, double-extended — of floating-point numbers. Each format can represent NaNs (Not-a-Number), $\pm\infty$ (infinity), and its own set of finite real numbers all of the simple form $2^{k+1-N}n$ with two integers n (signed significant) and k (unbiased signed exponent) that run throughout two intervals

$$1 - 2^K < k < 2^K \quad (K + 1 \text{ exponent bits}) \text{ and}$$

$$-2^N < n < 2^N \qquad (N \text{ significant bits})$$

determined from the format.

The IEEE standard specifies how single precision (32 bit) and double precision (64 bit) floating point numbers are to be represented. It also specifies how arithmetic should be carried out on them. The VMS front-ends, the Ultrix front-ends, and the Cray C90 systems use IEEE formats. The differences in the formats may affect the accuracy of floating point computations.

Single precision IEEE format The IEEE single precision floating point standard representation needs a 32 bit word, numbered from 0 to 31, left to right. The first (0th) bit is the sign bit, s, the next 8 bits are the exponent bits eeee eeee, and the final 23 bits are the fraction bits ffff ffff ffff ffff ffff fff (Figure 4.4).

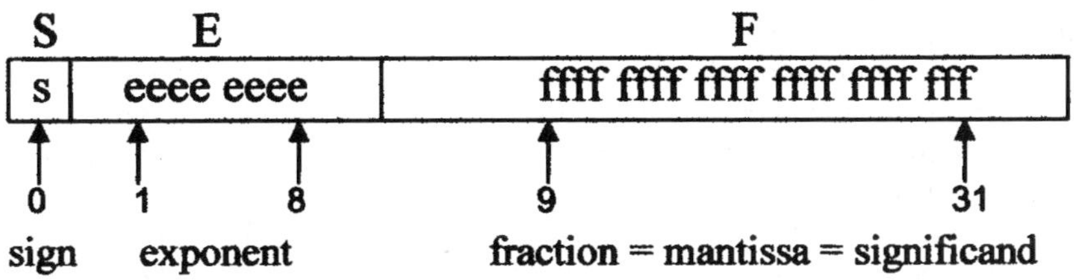

Figure 4.4: The IEEE single-precision (32 bit word) floating-point format

The actual value v represented by the contents of the word is obtained by the following rules, denoting **0** (bold) = 0000 and **1** (bold) = 1111:

a. If S=0 or 1 and E=$(\mathbf{11})_2$=$(255)_{10}$ and F≠0 then v=NaN (not-a-number).

b. If S=1 and E = $(\mathbf{11})_2 = (255)_{10}$ and F = 0 then v = −∝ (minus infinity).

c. If S=0 and E = $(\mathbf{11})_2 = (255)_{10}$ and F = 0 then v = +∝ (plus infinity).

d. If 0<E<$(255)_{10}$ then v = $1.F\times(-1)^S\times2^{E-127}$,
where 1.F represents the binary number created by prefixing F with the *implicit leading* 1 and a binary point.

e.. If E=$(\mathbf{00})_2 = (0)_{10}$ and F ≠ 0 then v = $0.F\times(-1)^S\times2^{-126}$ (unnormalized).

f. If S=1 and E=$(\mathbf{00})_2 = (0)_{10}$ and F = $(\mathbf{00000}\ 000)_2$=$(0)_{10}$ then v= −0.

g. If S=0 and E=$(\mathbf{00})_2 = (0)_{10}$ and F = $(\mathbf{00000}\ 000)_2$=$(0)_{10}$ then v= +0.

Thus, the single precision (32 bit word) representation for some numbers is

0 **00** **00000** 000 = +0

1 **00** **00000** 000 = −0

0 **11** **00000** 000 = +∝

1 **11** **00000** 000 = −∝

0 **11** **0** 01 **0000** 0 = NaN

0 **11** 0010 1010 0010 0101 0101 010 = NaN

0 1000 **0** **00000** 000 = $+1.0 \times 1 \times 2^{128-127} = 2$ (in excess 127 code)

0 1000 0001 1100 **0000** 000=$(1.11)_2\times(+1)\times2^{129-127} = (1+.5+.25)\times2^2$=7

1 1000 0001 1100 **0000** 000=$(1.11)_2\times(-1)\times2^{129-127}$ = −7

0 **0** 0010 **00000** 000 = $1.0\times(+1)\times2^{2-127}$=$2^{-125}$=2.3509887016e−038
(< machine epsilon but > dwarf in 32 bit (single) precision)

0 **00** 1000 **0000** 000=$(0.1)_2\times(+1)\times2^{-126}$=$2^{-127}$.
(Unnormalized number, Rule e)

0 **00** **00000** 001=$2^{-23}\times(+1)\times2^{-126}$=$2^{-149}$ = 1.4012984643e−045
(Unnormalized number, Rule e, smallest positive number=dwarf in single (32 bit) precision)

4.9.1 Double precision IEEE Format

The IEEE double precision floating point standard representation needs a 64 bit word, which may be represented as numbered from 0 to 63, left to right. The first bit is the sign bit, S, the next eleven bits are the exponent bits, 'E', and the final 52 bits are the fraction 'F':

The IEEE double precision floating point standard representation needs a 64 bit word, numbered from 0 to 63, left to right. The first (0th) bit is the sign bit, s, the next 11 bits are the exponent bits eeee eeee eee, and the final 52 bits are the fraction bits ffff ffff ffff ffff ffff ffff ffff ffff ffff ffff ffff ffff ffff (Figure 4.5).

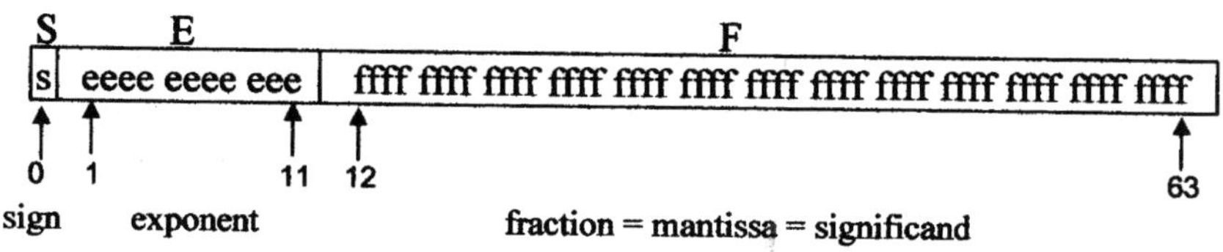

Figure 4.5: The IEEE double-precision (64 bit word) floating-point format

The actual value v represented by the contents of the word is obtained by the following rules, denoting **0** (bold) = 0000 and **1** (bold) = 1111:

a. If S=0 or 1 and E=(**11** 111)$_2$=$(2047)_{10}$ and F$\neq$0 then v=NaN (not-a-number).

b. If S=1 and E = (**11** 111)$_2$ = $(2047)_{10}$ and F = 0 then v = $-\infty$ (minus infinity).

c. If S=0 and E = (**11** 111)$_2$ = $(2047)_{10}$ and F = 0 then v = $+\infty$ (plus infinity).

d. If $0<E<(2047)_{10}$ then $v = 1.F\times(-1)^S\times2^{E-1023}$,

where 1.F represents the binary number created by prefixing F with the *implicit leading* 1 and a binary point.

e. If E=(**00**)$_2$ = $(0)_{10}$ and F $\neq$ 0 then $v = 0.F\times(-1)^S\times2^{-1022}$ (unnormalized).

f. If S=1 and E=(**00**)$_2$ = $(0)_{10}$ and F = (**00000** 000)$_2$=$(0)_{10}$ then v= –0.

g. If S=0 and E=(**00**)$_2$ = $(0)_{10}$ and F = (**00000** 000)$_2$=$(0)_{10}$ then v= +0.

It can be seen that the IEEE standard represents single precision (32 bit) exponents in excess127 code and not in excess 128 code (ieee.html). It permits representation of unnormalized numbers in addition to normalized ones.

4.9.2 The IEEE 854 standard

Besides IEEE 754 which is a binary standard, there is the IEEE 854 standard that permits the radix $\beta = 2$ or $\beta = 10$. The base $\beta = 10$ represents the conventional number system used and thoroughly understood by humans all over the world and is specially suitable for calculators). Unlike IEEE 754, it does not specify how floating point numbers are encoded into bits (Cody et al. 1984). The 854 standard specifies constraints on allowable values of the precision p for the single precision as well as the double precision but it does not need a particular value for p.

4.9.2.1 Why standard

Are the IEEE standards the best? Any standard/format will not be accepted as the standard if it is not *a best*. We do need a standard (to be widely followed) for several reasons. If we use different floating point formats for different machines then portability of programs could be a problem; in addition there would be problems for number representation, computation, and interfacing. Do we really have a better standard than the IEEE standards rather globally useful and acceptable under the various restrictions (constraints)? The answer is not emphatically no nor emphatically yes. Under various implementations with certain amount of generality, the IEEE standards seem excellent and possibly a best as of now (2003). The improvement on these standards could replace the current ones in future and we might have a better standard. For an introduction to the IEEE standards and their use, refer Cody (1988) and IEEE (1987).

4.9.3 Lower radix with higher precision versus higher radix with lower precision

Let the radix (i) $\beta = 2$ and the precision $p = 4$ and (ii) $\beta = 16$ and $p = 1$. Both the number systems have 4 bits of mantissa, Consider the computation of 13/8. If $\beta = 2$ then 13 is represented as 1.101×2^3 and 13/8 as 1.101×2^0. Thus the representation of 13/8 is exact in system (i). In system (ii), 13 is represented as $D \times 16^0$, where D is the hexadecimal representation of the decimal number 13. The rational number 13/8 is represented as 1×16^0, which has only 1 digit correct. Take, as another example, the rational number 29/16 with (i) $\beta = 2$ and $p = 5$ and (ii) $\beta = 16$ and $p = 2$. In system (i), 29/16 is represented as $1.1101 \times 2^4/2^4 = 1.1101 \times 2^0$ while in system (ii) it is represented as $1D \times 16^0/16^1 = 1.D \times 16^0$ which has both the digits correct; the representation is exact in both the system. However, base 16 can lose maximum 3 bits as shown in the first example. So a precision p in base 16 can have an effective precision of 4p – 3 bits instead of 4p bits. Similarly base 32 can have an effective precision of 5p – 4 bits instead of 5p bits.

4.9.3.1 Base 2 is preferable to base 16

Since large values of a base has reduced effective precision, what could be the reasons for IBM to choose base 16 for its system/370 during 1970s? One reason could be *increased exponent range*. In the IBM system/370 machine (which is 32 bit word) in the single precision (i.e., 32 bits), p = 6 hexadecimal digits. Hence the mantissa needs 6 × 4 = 24 bits but the effective precision can be as low as 4p – 3 = 21 bits. That is, the range of precision is 21-24 bits in the system/370 machines in the single precision. Out of the remaining 8 bits, 1 bit is used for sign and 7 bits for exponent (in excess code). Thus the magnitude of representable number in system/370 machines ranges from about 16^{-64} to about $16^{64} = 2^{256} = 1.1579 \times 10^{77}$.

To get a similar exponent range when the base β is 2, we would need 9 bits of exponent leaving only 22 bits for mantissa (as 1 bit is always for the sign of the mantissa). Another reason to choose β = 16 could be *less number of shifting* operations. Consider, for example, the addition of two floating point numbers with different exponents. One of the mantissas (usually the mantissa of the smaller exponent) needs to be shifted to make both the exponent same i.e., to make the radix point line up. This shifting slows down the add operation. In the β = 16 and p = 1 system all the numbers in [1, 15] have the same exponent. Hence no shifting is needed when adding $^{15}C_2 = 105$ possible pairs of distinct numbers

(1, 14), (2, 13), (3, 12), . . ., (14, 1); ←	14 pairs
(1, 13), (2, 12), (3, 11), . . ., (13, 1); ←	13 pairs
(1, 12), (2, 11), (3, 10), . . ., (12, 1); ←	12 pairs
(1, 4), (2, 3), (3, 2), (4, 1); ←	4 pairs
(1, 3), (2, 2), (3, 1); ←	3 pairs
(1, 2), (2, 1); ←	2 pairs
(1, 1). ←	1 pair

Observe that pairs (1, 14) and (14, 1) are two different pairs. In the β = 2 and p = 4 system, the foregoing numbers have exponents in [0, 3] and shifting is required for 70 out of 105 pairs. However, the gain by avoiding shifts in 21st century computers is too small. *Hence base β = 2 with small (rather smallest) wobble is the preferable base.*

It has yet another advantage. In β = 2 system, it is possible to gain one extra bit of significance in the mantissa (Goldberg 1967; Knuth 1981). This is because of the fact that floating point numbers are always normalized, the most significant bit of the mantissa is always 1. Consequently we need not have to retain a bit of storage in hardware for this 1. Formats employing this convention saving 1 bit are assumed to have a *hidden bit*. By making use of the hidden bit, IEEE 754 single precision has the mantissa 24 bits, i.e., p = 24 although only 23 bits are actually used.

4.9.4 Four different precisions of IEEE 754 standard

Besides the usual single and double precisions, the IEEE standard defines two more precisions, viz., single extended and double extended. In the single extended format, the precision (mantissa) $p \geq 32$ (as against p = 24 in single precision) while in the double extended format, $p \geq 64$ (as against p = 53 in double precision). The exponent widths in single and double precisions are 8 and 11 bits, respectively while these in the single extended and double extended precisions are ≥11 and ≥15 bits, respectively. The lengths of the floating point number are 32, 64, ≥43, and ≥79 in single, double, single extended, and double extended formats, respectively. The minimum permissible length of double extended format is 80 bits instead of 79 bits since this format does not make use of the hidden bit. The range of exponent in the single precision format in IEEE 754 is [–126, 127] while in the single extended format, the minimum exponent ≤ –1022 and the maximum exponent ≥ 1023. The range of exponent in the double precision format is [–1022, 1023] and in the double extended format, the lowest exponent ≤ –16382 and the largest exponent ≥ 16383.

4.9.4.1 Motivation for extended precision

The extended precision permits computation of quantities to within 0.5 ulp (units in the last place — defined later) in single/double precision efficiently. A single precision number will be printed with enough decimal digits so that when the decimal number is read back in, the single precision number may be recovered. A decimal number with 9 decimal digits is sufficient to recover a single precision binary number in IEEE format (Goldberg 1991). While converting a decimal number back to its unique binary number, even a small error of 1 ulp could produce a wrong result. The extended precision is useful to avoid such a fatal error. To obtain a good quality result, an algorithm should use extended precision. An expression that approximates log with an absolute error of 500 ulp will produce correct result up to 10 digits when computations are carried out with 13 digits. The extra three digits are kept hidden from the user/viewer.

4.9.5 Computation using IEEE floating point standard

The result of a basic arithmetic operation in IEEE standard is computed exactly and then rounded, using round-to-even, to the nearest floating point number. If the exponents of two floating point numbers are widely different then addition/subtraction of these numbers could be expensive and there could be a loss of significance. The output of an arithmetic operation where a single guard digit has been used may differ from the output obtained by computing the exact result followed by rounding. Using two guard digits along with a *sticky* bit (Goldberg 1990) such a difference vanishes.

IEEE standard specifies the output of arithmetic computations precisely and completely to enhance portability of software as well as ease in reasoning about floating point computation. When a computer program is shifted from one machine to another, where both the machines support IEEE standard, the final output will remain the same. If it is found that some intermediate results differ, then it would be due to bugs in the software and not due to the arithmetic. For axioms (self-evident statements accepted without proof) on floating point number and arithmetic, see Brown (1981).

Operations covered by IEEE arithmetic The IEEE standard specifies that, besides the four basic arithmetic operations, remainder, square-root, and conversion between integers and floating point numbers as well as between internal formats and decimals be correctly rounded. This specification does not include very large numbers.

The computation of an inner product in IEEE arithmetic could be incorrect. For example, the inner product of the two vectors $a = [2 \;\; 10^{30} \;\; 1 \;\; 1]^t$ and $b = [\, 10^{-30} \;\; 1 \;\; -10^{-30} \;\; -10^{30}]^t$ the errorfree value of the inner product ab $= ((2 \times 10^{-30} + 10^{30}) - 10^{-30}) - 10^{30}$ is 10^{-30}, where the parentheses denote the order in which the arithmetic is done. But the computer with IEEE arithmetic produces the result $-\ 10^{-30}$ (Goldberg 1991). Because of this problem Kulisch and Miranker (1986) suggested the inclusion of the inner product to the list of operations that are precisely specified. Kirchner and Kulisch (1987) showed that the inner product can be computed within 1 ulp with less hardware than that required to implement a fast multiplier. All the IEEE arithmetic operations excluding the decimal-binary conversion need to be exactly rounded as efficient algorithms for exactly rounding all nonconversion operations are available. For conversion, however, the best algorithms produce nearly the same results as the exactly rounded ones (Coonen 1984).

The IEEE arithmetic does not need transcendental functions to be exactly rounded. This is because of non-practicability of specifying the precision of the transcendental functions to be the same as if these are computed to infinite precision and then rounded (Goldberg 1991).

4.9.6 Does any bit pattern represent a valid floating point number?

On IBM system/370 floating point hardware, any bit pattern of a 32 bit word represents a valid floating point number. This is, however, not the case with CDC 6600 computer in which certain bit patterns represent special quantities such as *infinity* and *indefinite*. Digital Equipment Corporation's VAX system also has certain bit patterns that represent *reserved operands*.

The IEEE standard like the CDC and the VAX systems represents certain bit patterns as information (useful in floating point computation) different from valid normalized floating point numbers. Some of these patterns represent NaN's (not a number), $\pm \infty$, denormalized numbers, and $\pm$ 0. There are several NaN's. These special patterns are all encoded with exponent of $e_{max} + 1$ or $e_{min} - 1$. These exponent values are obviously outside the exponent range $[e_{min}, e_{max}]$.

In IEEE 754, NaN's are represented as floating point numbers with nonzero mantissas and with exponent $e_{max} + 1$. In IEEE arithmetic, the square root of a negative number, say – 25, returns a[12] NaN as output while the IBM system/370 Fortran (existing during early 1970's, now obsolete, and possibly not available anywhere for computation) produces 5 — a wrong output (a valid floating point number though).

4.9.7 Role of NaN

Division by 0, 0 divided by 0, square-root of a negative number are operations which cause unrecoverable error and consequently traditionally halt the computation. This halting is not desirable in several situations. Consider, for example, a fixed point iteration scheme to compute roots of an equation in one or more variables. It might so happen that for one of the several roots, the scheme could come across 0/0 or a division of a finite nonzero number by 0 or square-root of a negative number. In such a situation, the computation would halt undesirably aborting the root-finding process. Consider the equation $x + \sqrt{(x^2 - 3x)} - 1 = 0$ which has one root at $x = -1$. If we use the fixed point iteration scheme

$$x_{i+1} = 1 - \sqrt{(x_i^2 - 3x_i)},\ i = 0, 1, \ldots, |x_{i+1} - x_i|/|x_{i+1}| \le .5 \times 10^{-4},$$

[12] The article "a" will be used if we pronounce NaN as *nan*. The article "an" will be used if we pronounce NaN as *en-a-*n just as we write "an n × n matrix" and not "a n × n matrix". We prefer to use "a NaN" rather than "an NaN".

where x_0 is an initial approximation. If we choose $x_0 = 1$ then the computation of x_1 would come across the computation of the square-root of -2 — an unrecoverable error. Consequently the root-finding scheme will be aborted/halted. If we now consider the equation $(x^4 - 1)(x + \sqrt{(x^2 - 3x)} - 1) = 0$ which has $x = 1$ and $x = -1$ as two of the roots and use the scheme

$$x_{i+1} = x_i^4(x_i + \sqrt{(x_i^2 - 3x_i)} - 1) + 1 - \sqrt{(x_i^2 - 3x_i)},$$

$$i = 0, 1, \ldots, |x_{i+1} - x_i|/|x_{i+1}| \le .5 \times 10^{-4},$$

where x_0 is an initial approximation. If we choose $x_0 = 1$ then, as before, the computation of x_1 would come across the computation of the square-root of -2 — an unrecoverable error although $x = 1$ is a root. The problem of halting can be obviated by introducing NaN that will be produced in the arithmetic operations $\infty + (-\infty)$, $0 \times \infty$, $0/0$, ∞/∞, ∞ rem x, x rem 0, and $\sqrt{(-x)}$, where x is a valid positive finite floating point number. When roots of an equation $f(x) = 0$ are attempted to be computed outside the domain of f due to possibly some incorrect initial approximation (guess), the execution of the subprogram for f will return a NaN without stopping the root-finding process. If a NaN is an operand in a floating point operation, then the output will be a NaN. If the value of a subexpression is a NaN, then the value of the entire expression will be a NaN.

4.9.8 Role of infinity

Like NaN's, ∞'s provide a way to continue the process of computation when a floating point overflow occurs. The use of ∞ is more desirable than allowing the overflow replaced by the largest representable number (in magnitude). Consider, for example, computing $\sqrt{(a^3 - b^3)}$, where the base $\beta = 10$, the precision $p = 4$, the largest value of the exponent $e_{max} = 85$, $a = 3 \times 10^{40}$, $b = 2 \times 10^{40}$. Then $a^3 = 27 \times 10^{120}$ will overflow and is replaced by 9.999×10^{85}. Similarly b^3 as well as $a^3 - b^3$ will both overflow and each is replaced by 9.999×10^{85}. So the final result will be $\sqrt{(9.999 \times 10^{85})} = 9.999 \times 10^{42}$ which is unacceptably wrong. The correct answer is 4.359×10^{60}. In IEEE standard, a^3 is ∞, so are $b^3$and $a^3 - b^3$. The final result $\sqrt{(a^3 - b^3)}$ is ∞. This is safer than getting a valid floating point number which is completely wrong.

A nonzero positive number a divided by 0 results in infinity: $a/0 = \infty$, $-a/0 = -\infty$. Consider, for $a > 0$, the computation of the function $g(x)/h(x)$ for $x \to 0$ in the limit, where $g(x) \to a$ and $h(x) \to 0$ from the positive side. then $g(x)/h(x)$ returns $+\infty$. If $a < 0$, then it returns $-\infty$. Observe that, for 0/0 or ∞/∞ situation, a NaN is returned. Thus the IEEE arithmetic defines $a/0 = \pm\infty$ as long as $a \ne 0$. The sign of ∞ depends on the sign of a and the sign of 0 in the usual mathematical sense. For a finite positive number a, the operation a/∝ returns 0, a − ∝ returns −∝, and a + ∝ returns ∝. The operation √∝ returns ∝. Let $k = e_{max} + 1$. The infinity arithmetic returns a wrong value 0 for the expression $x/(x^4 - 2)$ when $x > \beta^{k/4}$ (as x^4 overflows) and not a value near $1/x^3$. A better expression for the infinity arithmetic is $1/(x^3 - 2\times x^{-1})$ since this expression will not overflow prematurely as the arithmetic will produce correct value 0 for x = 0. If we do not have the infinity arithmetic, then the expression needs a test whether x = 0. This "test" is an additional instruction for sequential machines. For a pipeline machine, this instruction is not only extra but also is undesirable as it could disrupt a pipeline.

4.9.9 Role of signed zero

In IEEE standard, 0 is represented with mantissa 0 and exponent $e_{min} - 1$. If the sign bit is 1, then it will represent −0 else +0. The standard defines the comparison so that +0 = −0 (and not −0 < +0). The only disadvantage of this comparison is that the relation a = b ⇔ 1/a = 1/b will not hold when a = +0 and b = −0. The standard does not ignore the sign of zero although it could be done so always. If the zero did not have a sign then the relation 1/(1/a) = a would fail to hold for a = ±∝. If we have only one kind of ∝ then we would lose the sign information of an overflowed number which could result in drastically wrong result.

The signed zero has another use. It is in (i) underflow, and (ii) discountinuous function f(x) at x = 0. In IEEE standard, $\log_e 0$ returns −∝ while log a returns a NaN for a < 0. If the number a happens to be such a small negative number that it underflows, then log would return a NaN and not −∝.

Yet another use of signed zero is in complex arithmetic. Consider the identity (Goldberg 1991) √(1/a) =1/√a. If a = −1, then √(1/a) = i while 1/√a = −i. Thus the identity is not satisfied! This mathematical fallacy lies in the fact that square-rooting is a two-valued function and there is no way to choose the values so that these are continuous in the entire complex plane. The problem with the negative real numbers of the form −a + i0, where a ≥ 0 can be resolved perfectly using the signed zero. If a = −1 = −1 + i0, then 1/a = −1 + i(−0). Hence √(1/a) = √(−1 + i(−0)) = −i since √(−a + i0) = i√a and √(−a −i0) = −i√a on the other side of the branch cut (Kahan 1987). Also, 1/√a = −i.

4.9.10 Denormalized numbers

The IEEE standard uses denormalized[13] numbers or denormals (Goldberg 1967) which ensure that $a = b \Rightarrow a - b = 0$ and satisfy other useful relations. If the bits in the mantissa after the leftmost digit are $b_1, b_2, \ldots, b_{p-1}$ and the value of the exponent is $e = e_{min} - 1$, the floating point number being represented is $0.b_1b_2 \ldots b_{p-1} \times 2^{e+1}$. If $e > e_{min} - 1$, then the floating point number being represented is $1.b_1b_2 \ldots b_{p-1} \times 2^e$. Most hardware computers that claim to be IEEE compatible support denormalized numbers only indirectly by software traps and not through hardware when accepting or generating denormals. These computers leave the task of simulating the IEEE standard to software. Consequently, a program producing underflows too often will run slower on the computers that use such software traps. If we know that the implication $a = b \Rightarrow a - b = 0$ is true, then writing dependable floating point code is simpler just as the knowledge of the implication $a/2 < b < 2a \Rightarrow a \ominus b = a - b$ is helpful. The floating point numbers need not be normalized when the exponent is e_{min}. When $\beta = 10$, $p = 4$, and $e_{min} = -98$, the floating point number 1.000×10^{-98} is not the smallest number. A still smaller floating point number is 0.789×10^{-98}. With the foregoing β, p, and exponent, if $a = 8.712 \times 10^{-97}$ and $b = 8.711 \times 10^{-97}$, then $a - b$ does not become zero. Instead, it results in the denormal 0.010×10^{-98}. This behaviour is known as *gradual underflow* (Demmel 1984; Goldberg 1991) which is desirable. A simple arithmetic expression $a + b$ may have a large relative error for normalized inputs. Observe the gulf (gap) between 0 and the smallest normalized number $1.0 \times \beta^e$ (say, $= 1.000 \times 10^{-98}$), where $e = e^{min}$ ($= -98$). During a floating point computation, if a number falls into this gap and, instead of being flushed to zero, is represented by the nearest denormal, then the gap is filled in when such a denormal is included in the number line. The adjacent spacings are the same length or differ by a factor of β. If denormals are not used, then the spacing all of a sudden changes from β^{e-p+1} to β^e which is a factor of β^{p-1} (where $e = e_{min}$), instead of a gradual change by a factor of β. Let $d/c = E$, $D = c + dE$, $G = c/d$, $H = d + cG$. Then the complex division $(a + ib)/(c + id)$ using Smith's formula can be written as

$(a + ib)/(c + id) = (a + bE)/D + i(b - aE)/D$ if $|d| < |c|$ else

$(a + ib)/(c + id) = (b + aG)/H + i(-a + bG)/H$.

Consider the following example (Goldberg 1991). If $a = 2 \times 10^{-98}$, $b = 10^{-98}$, $c = 4 \times 10^{-98}$, and $d = 2 \times 10^{-98}$, then the complex division, applying Smith's formula, returns the correct answer .5 with gradual underflow while, with flush to zero, it returns the value .4 that has an absolute error of 100 ulps.

---

[13] Called subnormals in IEEE 854 standard

However, in double/double-extended precision (e.g., in Matlab), we get the correct answer .5.

For further information on exceptional conditions (division by zero, overflow, NaN for 0/0 and $\sqrt{-1}$, and $\propto$ for 1/0), the resulting status flag, and trap handlers, refer Goldberg (1991).

4.9.11 Compare floating point numbers for equality

Two equal floating point numbers when compared for equality could produce the answer “not equal’ depending on the design decisions. For example, the Boreland’s Turbo Basic on an IBM personal computer had returned “not equal” for exactly equal numbers (Goldberg 1991). The design decisions are made taking into account such an eventuality which is not so uncommon. For instance, the Matlab program segment

```
» a=5.0/11.0;if (a==5.0/11.0), result='equal'; else, result ='not
equal';end;result
```

will return

```
result = equal
```

which is correct and desirable. A similar program using the foregoing Turbo Basic prints “not equal”.

4.9.12 Kahan’s summation formula: Improvement over simple formula

Kahan’s formula for the sum $s = \Sigma a_i$ $(i = 1, 2, \ldots, n)$ is given by the following Matlab commands.

```
»n=1000; for i=1:1000, a(i)=i; end; z=0; s=a(1); for i=2:n, b=a(i)-z;
u=s+b; z=(u-s)-b;s=u; end;
» s
```

The computed sum s has an error $\leq 2\varepsilon\Sigma a_i$ neglecting the terms $O(n\varepsilon^2)$. The simple formula produces the sum s with an error $\leq n\varepsilon\Sigma a_i$ (Goldberg 1991), where ε is the machine epsilon. Observe that the floating point arithmetic need not obey the algebraic identity $z = ((s + b) - s) - b = 0$. The code optimiser should, therefore, be cautious while applying algebraic identities that are true for real numbers to expressions involving floating point variables/numbers.

4.9.13 Failures in comparisons of two floating point variables

The Boolean constant *true* cannot replace a = a since it fails when x is a NaN. When a = +0, –a = 0 – a fails. Further a < b is not the complementary inequality of a ≥ b. All these are due to the fact that NaNs cannot be compared with ordinary floating point numbers.

Some of the algebraic identities that are valid for floating point numbers in IEEE arithmetic are 1 × a = a, .5 × a = a/2, 2 × a = a + a, and a + b = b + a. However, even these identities may fail on a few machines such as CDC and Cray supercomputers (Goldberg 1991).

4.9.14 Rounding modes in IEEE with application to interval arithmetic

Except binary-decimal conversion, each operation is computed exactly and then rounded in IEEE arithmetic. Besides *round toward nearest* which is the default meaning of rounding, the arithmetic needs that three additional modes, i.e., *round toward –∝* (floor operation in the context of integer arithmetic), *round toward 0,* and *round toward ∝* (ceiling operation in the context of integer arithmetic) be included. In the round toward 0 mode or in the round toward –∝ mode, an overflow of positive magnitude produces the largest positive number but not +∝. In the same way an overflow of negative magnitude will cause the default result to be the largest negative number but not –∝ in the round toward +∝ mode or in the round toward 0 mode.

In the interval arithmetic (Moore 1961; Gibb 1961; Chartres 1966; Rokne and Lancaster 1971) each number is represented as an interval. If two intervals $[a_d, a_u]$ and $[b_d, b_u]$ are added then the result is $[c_d, c_u]$, where $c_d = a_d \oplus b_d$ with the rounding mode set to round toward –∝, and $c_u = a_u \oplus b_u$ with the rounding mode set to round toward +∝. In floating point computation using the interval arithmetic, the result is produced as an interval which contains the exact output (never known) of the computation. If the interval (output) is large (which is often the situation) then the interval arithmetic is not very useful. The interval arithmetic could be more useful if it is used with multiple precision floating point arithmetic.

4.9.15 Measuring rounding error in floating point computation

If the result of a floating point computation in the finite precision p = 3 and in base β = 10 is 5.17×10^{-3} and the answer when computed with infinite precision is .00519, the result is in error by 2 units in the last place. If the result is 7.13×10^{-2} while the actual real number is .0713186 then the error is by .186 in the last decimal place. In general, if the finite precision p floating point number is a.ff . . . f × β^E for the real number r then this floating point number is in error by $|a.ff\ldots f - (r/\beta^E)|\ \beta^{p-1}$ *units in the last place* (ulps). The ulps is an estimate of the *absolute error* in the number. In a floating point computation with floating point, if the result is nearest the true value then this result could still have maximum ½ ulp error.

Another way to measure the difference of a floating point number and the true real number is to compute an estimate of relative error which is the foregoing difference (in magnitude) divided by the true (higher order accurate) real number. The relative error, for instance, committed when approximating .0713186 by 7.13×10^{-2}, is $|.0713186 - 7.13 \times 10^{-2}|/|.0713186| =$ 2.608015300355534e-004. The relative error bounds can be written as $[\beta^{-p}, \beta^{-p+1}/2]$, where p = the precision of computation. These bounds bracket the value ulp/2. The relative error bounds containing ulp/2 can vary by a factor of $\beta/2$. Twice the factor, viz., β is known as *wobble* while the larger of the two bounds $\in = \beta^{-p+1}/2$ is called the machine epsilon. *The machine epsilon* $\in$ *implies that when a real number is rounded to the nearest floating point number, the relative error is bound by* $\in$*, i.e., the relative error bounds will be* $[\beta^{-p}, \in]$ *for the closest floating point number.* The relative error may often be expressed in terms of $\in$. In the foregoing numerical example, the relative error 2.608015300355534e-004 can be written as $(2.608015300355534\text{e-}004/(\beta^{-p+1}/2)) \times \in = 0.05216030600711\in \approx 0.05\in$ when $p = 3$ and $\beta = 10$. This representation is just to avoid writing very small number.

4.9.16 Difference between ulps and relative error

If a real number $r = 17.38$ is represented by $r' = 1.74 \times 10^1$, where $p = 3$ and $\beta = 10$ and consequently $\in = 5 \times 10^{-3} = 0.005$, then the absolute error estimate is $|r - r'| = 0.02 = 0.2$ ulps while the relative error is $|r - r'|/|r| = 0.0012 \approx 0.23\in$. If we now compute $9 \times r$ then we obtain 156.42 while the computed value is $9 \times r' = 1.57 \times 10^2$. The absolute error is now 5.8 ulps but the relative error $= |9r - 9r'|/|9r| = 0.0012 \approx 0.23\in$ remains unaltered. Thus, the absolute error measured in ulps is $5.8/0.2 = 29$ times larger while the relative error remains unchanged. The reader may refer Chapter 2 to see that this is precisely what it should be with respect to the absolute and the relative errors. A fixed absolute error of 0.5 ulps may result in a relative error that can wobble (move unsteadily) by β (the radix value). On the contrary, a fixed relative error expressed in ulps can wobble by a factor $\le \beta$.

4.9.16.1 Ulps versus relative errors: scope

The rounding error in the nearest floating point number for a real number is (maximum) 0.5 ulps always. Thus it is natural and convenient to express rounding errors in floating point numbers in terms of ulps. However, when analysing the rounding errors in computation, the relative error is a better measure as it provides us the relative information among the computed quantities, unlike the absolute error estimate in ulps. The ulps or the machine epsilon $\in$ may be used when we wish to determine the order of rounding error (in modulus) in a floating point number. This is because $\in$ and ulps differ by a factor of maximum β. If a floating point number is in error by k ulps then the number of contaminated digits is $\log_\beta k$. If the relative error in a computation is $k\in$ then the number of contaminated digits is $\log_\beta k$.

For further details refer the website

http://www.cs.berkeley.edu/~wkahan/ieee754status/IEEE754.PDF

as well as other concerned websites on IEEE standard 754 for binary floating-point arithmetic.

4.9.17 Why floating point numbers have to be rounded

Squeezing infinitely many real numbers in a finite number of bits is impossible. The best that can be done is to represent these numbers approximately. Given a fixed number of bits, say 32, most computations with real numbers (which are essentially rational numbers, rather a small fraction of rational numbers) will produce real numbers that cannot be exactly represented using that many bits. Thus, the output of a floating-point operation must often be rounded to fit back into its finite representation.

4.9.18 Why guard digit

If add, subtract, multiply, and divide operations introduce a bit more rounding error than necessary then this bit provides a means of reducing the rounding error when subtracting two nearly equal numbers. This bit is called a *guard digit*. In late 1960's IBM considered the introduction of guard digits important. Consequently it added a guard digit in both the single and double precision formats in IBM 360 computers. The IEEE standard uses a guard digit and, in addition, provides an algorithm for the basic arithmetic operations and for square-rooting. The implementation of this algorithm is such that it produces the same result (in every bit) in all the machines that support the IEEE standard. This implementation simplifies portability of programs from one machine to another.

We state the following theorems due to Goldberg (1991) without proof to show the importance of introducing a guard digit in reducing the rounding error. Assume that the floating point format has the precision p and the base β.

Theorem 1 (*Relative error without a guard digit*) The relative rounding error in computing the difference of two floating point numbers can be as large as $\beta - 1$ (large relative error) under the foregoing assumption.

Theorem 2 (*Relative error with a guard* digit) The relative rounding error in subtracting (or adding with opposite signs) two floating point numbers will be less than $2\in$ (small relative error) if the subtraction is done with p + 1 digits (i.e., p digits plus one guard digit) under the foregoing assumption.

If we do not use a guard digit then the relative error in subtracting two nearly equal (nearby) floating point numbers could be significantly large (Theorem 1). It could be even so large that all the digits in the result could be contaminated — this is due to *catastrophic cancellation* (Goldberg 1991).

When subtracting two rounding errorless quantities using a guard digit then the result has a small relative rounding error $<2\in$ (Theorem 2). This is due to *benign cancellation* (Goldberg 1991). Most recent computers have a guard digit; there are a few (e.g. the Cray supercomputer systems) which do not use a guard digit.

Consider as an example, the computation of the area A of a triangle whose sides are a, b, and c in lengths. If the perimeter $2s = a + b + c$ then

$$A = \sqrt{[s(s - a)(s - b)\ (s - c)]}.$$

If A is very small, i.e., the triangle is nearly needle-like or, equivalently, nearly flat then $a \approx b + c$, i.e., $s \approx a$. Then the computation $s - a$ will involve large rounding error if a guard digit is not used. If we rewrite A, where $a \geq b \geq c$, $c1 = a - b$, and $q = b - c$, as (Kahan 1986)

$$A = \sqrt{[2s(c^2 - c1^2)(a+q)]/4}$$

then we may get more accurate result for a flat (almost like a straight line) triangle. Observe, in this context, that to obtain $a - b$, where $a \approx b$, we might get better result if we compute $(a^2 - b^2)/(a + b)$ without a guard digit. We now state Theorems 3 and 4 (Goldberg 1991).

Theorem 3 (*One-guard-digit based subtraction with .5 ulp square-rooting for a triangle*) The rounding error in computing the area of the triangle A using the formula $A = \sqrt{[2s(c^2 - c1^2)(a+q)]/4}$ is less than or equal to $11\in$ if the subtraction is performed with a guard digit, $\in \leq 0.005$, and square-root is computed within 0.5ulp.

Theorem 4 (*One-guard-digit-based subtraction with .5 ulp LN for ln*) Let $\oplus$ denote the computed addition. Assume that LN(x) approximates ln(x) to less than or equal to 0.5 ulp. If $\ln(1 + x)$ is computed using the formula, where $x1 = x + 1$,

$$\ln(1+x) = x(\ln x1)/(x1-1) \text{ if } 1 \oplus x \neq 1, \text{ else } x$$

then the relative error is less than or equal to $5\in$ for $0 \leq x < 0.75$ provided the subtraction is performed with a guard digit and $\in < 0.1$.

The foregoing formula is interesting for x much less than 1, where catastrophic cancellation occurs, although it will work for any value of x.

Exactly rounded operations If a floating point operation is performed with a guard digit then the operation is not as accurate as if it is computed exactly and then rounded to the nearest floating point number. A floating point operation performed in this manner is termed as an exactly rounded operation. For further details, refer Goldberg (1991).

4.9.19 Round-up and round-to-even operations: Which is better ?

There are two ways of rounding — (i) round-up and (ii) round-to-even. Both the ways are identical if the last digit to be rounded is not 5. In the round-up way, the ten decimal digits are equally divided — {0, 1, 2, 3, 4} and {5, 6, 7, 8, 9}. If the last digit ends with one of the digits of the set {0, 1, 2, 3, 4} then round down else round up. This is how the rounding works in VAX machines produced by Digital Equipment Corporation.

In the round-to-even way, if the last digit to be rounded to be 5 then round up if the preceding digit is odd else round down. Let a and b be two floating point numbers. Also, let $\oplus$ and $\ominus$ denote computed addition and subtraction (i.e., with rounding error). The following theorem (Reisser and Knuth 1975) then would demonstrate that round-to-even is better.

Theorem 5 (*Use of round-to-even*) Set $a_0 = a$. $a_1 = (a_0 \ominus b) \oplus b$, $a_2 = (a_1 \ominus b) \oplus b, \ldots, a_k = (a_{k-1} \ominus b) \oplus b$. Let the binary operations $\oplus$ and $\ominus$ be exactly rounded using the round-to-even way. Then $a_k = a$ for all k or $a_k = a_1$ for all k ≥ 1.

To illustrate Theorem 5, consider the decimal base $\beta = 10$, the precision $p = 4$, $a = 10.000$, and $b = -.5555$. If the round-up way is employed then $a_0 \ominus b = 10.556$. $a_1 = 10.556 \ominus .5555 = 10.001$, $a_2 = 10.002$, $a_3 = 10.003$, and so on. Thus each successive value of a_k increases by 0.001. If the round-to-even way is used then a_k is always 10.000 (by Theorem 5). From the foregoing numerical example, it is clear that, in the round up way, the successive results will climb up while, in the round-to-even way the successive results do not climb up or down (as it should be).

From the probability point of view, the use of round-up operations seems not that unjustified because each of the 10 digits 0, 1, 2, 3, 4, 5, 6, 7, 8, 9 is equally probable as the last digit in a computation. Situations such as the one in Theorem 5 are specific and not that general although still the use of round-to-even operations has an edge and is recommended.

In multiple precision arithmetic, exact rounding has an important application. There are two procedures of representing a multiple precision floating point number. In the first procedure, a floating point number with a large mantissa (fraction) is stored in an array of consecutive words and a program (routine) — usually assembly language — is written/used to manipulate the contents of these words.

In the second procedure, the multiple precision floating point number is represented as an array of ordinary floating point numbers, where the multiple precision floating point number is the addition of the contents (elements) of the array in the infinite precision. The second procedure is better in the sense that it can be programmed in a high level language portably subject to the use of exactly rounded arithmetic.

To compute the product ab, we split a and b, write $a = a_h + a_t$ and $b = b_h + b_t$ and then express ab as the sum $a_hb_h + a_hb_t + a_tb_h + a_tb_t$, where each summand has the same precision p even bits as that of a and b subject to the fact that each of a_h, a_t, b_h, and b_t can be represented using 0.5p bits. The p-digit number $a = a_0a_1a_2a_3 \ldots a_{p-1}$, where a_i is the i-th digit of the number, can be written as (the sum) $a = a_0a_1a_2a_3 \ldots a_{(0.5p)-1} + 0000 \ldots 0\, a_{0.5p}a_{(0.5p)+1} \ldots a_{p-1}$. This splitting will work for only even precision. For an odd precision, we may gain an extra digit by splitting a as the difference of two numbers. For instance, if the base $\beta = 10$, $p = 7$, and $a = .5055555$ then a can be split as $a = a_h - a_t = .51 - .0044445$. Out of several ways of splitting, the one that is easy to compute is due to Dekker (1971) subject to the use of 2 or more guard digits. We now can state the following result (Theorem 6), where the floating point precision p is even when $\beta > 2$, $k = \lceil 0.5p \rceil$ is half the precision (rounded up), $m = \beta^k + 1$, and the floating point operations are exactly rounded.

Theorem 6 (Splitting a number using exact rounding) The number a can be split as $a = a_h + a_t$, where $a_h = (m \otimes a) \ominus (m \otimes a \ominus a)$, $a_t = a \ominus a_h$, and each a_i can be represented using 0.5p digits of precision.

To illustrate Theorem 6, take the base $\beta = 10$, $p = 6$, $a = 4.46355$, $b = 4.47655$, and $c = 4.\ 47955$. Then $b^2 - ac = 0.0448045$ (rounded to the nearest floating point number), $b \otimes b = 20.03950$, $a \otimes c = 19.99470$ and so the computed value of $b^2 - ac = 0.0448$. This is in error of 045 ulps if we do not use Theorem 6. Using Theorem 6 we write $a = 4.5 - 0.03645$, $b = 4.5 - 0.02345$ and $c = 4.5 - 0.02045$. Hence $b^2 = 4.5^2 + 0.02345^2 - 2 \times 4.5 \times 0.02345 = 20.25 + .0005499025 - 0.21105$ which is not computed to a single value at this stage. Likewise, $ac = 4.5^2 + 0.03645 \times 0.02045 - (4.5 \times 0.02045 + 4.5 \times 0.03645) = 20.25 + .0007454025 - 0.25605$ (also not computed to a single value at this stage).

We subtract the foregoing two series term by term and get $0 \ominus 0.0001955 \oplus 0.045 = 0.0448045$ which is the exactly rounded value of $b^2 - ac$.

The fact that Theorem 6 needs exact rounding is illustrated by the following example. Consider $\beta = 2$ (binary), p = 3, and a = 5. Then $m = \beta^k + 1 = 2^2 + 1 = 5$, ma = 25 = $(11001)_2$, m ⊗ a = 24 (since p = 3). If the subtraction is performed with a single guard digit, then (m ⊗ a) Θ a = 20. Hence $a_h = 4$ and $a_t = 2$. Consequently, a_t is not representable with $\lfloor p/2 \rfloor = 1$ bit. Yet another instance, where the exact rounding is needed, is the computed division operation a ∅ 15. The floating point result will not be, in general, equal to a/15. For a binary number ($\beta = 2$), if we multiply (a ∅ 15) by 15, we will get back a provided exact rounding is used.

Theorem 7 (*Welcome effect of exact rounding on division and multiplication of integers*) Consider two integers a and b with $|a| < 2^{p-1}$ and b is of the form $b = 2^i + 2^j$. If the base $\beta = 2$ and floating point operations are exactly rounded, then (a ∅ b) ⊗ b = a.

The theorem holds true for any base β provided $b = \beta^i + \beta^j$. However, as β grows larger and larger, we will have fewer and fewer denominators b of the form $\beta^i + \beta^j$.

If the basic arithmetic operations, viz., add, subtract, multiply, and divide operations ⊕, Θ, ⊗, and ∅ produce slightly more rounding error than necessary, then this additional error, though small, could have significant effect on the final result. This is why several algorithms for the basic arithmetic operations use guard digits to perform accurate arithmetic computations that do not have the additional rounding errors. If the inputs a and b to these algorithms involve measurement errors (which is, in general, the case in almost all engineering problems), then the benign cancellation a − b may become catastrophic. Consequently, the importance of Theorems 3 and 4 might come down. Yet accurate arithmetic operations are useful even for inputs which are erroneous (inexact) due to imprecise measurement and approximate floating point representation (of actual value). This is because these algorithms for accurate arithmetic computations allow us to establish errorfree relationships such as those stated in Theorems 6 and 7.

The operation (e.g. scaling up or down by 2^k) that does not modify the fraction (mantissa) but changes the exponent which is integer does not produce any error. For the historical development and further information on errors in number representation and computation, refer Barnett (1987), Brown (1981), Cody (1988), Dekker (1971), Demmel (1984), Farnum (1988), Goldberg (1967, 1990, 1991), Golub and Van Loan (1989), IEEE (1987), Kahan (1972, 1987), Kahan and Coonen (1982), Kahan and LeBlanc (1985).

So far as the outside world is concerned, the numerical output of a computer is in decimal form easily understood by man in machine-man communication as well as in man-man communication. *Binary is hardly used in these communications*. The machine gets the correct (not necessarily exact) information in bit form through decimal-binary conversion (existing in the form of a firmware or a software program in the machine). The nonnumerical output (say, a message/text in English) easily understood by man is also usually in nonbinary form. The nonnumerical/alphanumeric input gets converted to a binary form (through a conversion routine) for the machine to process the information. Since everything inside the machine is in bits, there is a mechanism to tell the machine which bits (sequence of bits) represent numbers and which others (nonnumbers or alphanumeric and/or special characters or instructions).

Interval of doubt in fixed- and floating-point numbers and computation Rounding errors (using round-to-even rule or, equivalently, the best rule) may accumulate in fixed- and floating-point numbers. If only 7 digits are used to define the fixed-point number a = .1134000, then a would represent any real number x in the interval .1133995000 . . . ≤ x ≤ .1134005000 . . . . This interval of length 10^{-6} is called the interval of doubt for the fixed-point number a. Likewise the number b = .0011340 could be a rounded representation for any real number y in the interval (also of length 10^{-6}) .0011335000 . . . ≤ y ≤ .0011345000 . . . . A sum of intervals would be .1146330000 . . . ≤ z ≤ .1146350000 . . . so that the sum a + b could correspond to any real number z which has the interval of length 2×10^{-6}. Hence we see that the interval of doubt grows with each addition. The computed sum, however, might be quite close to the true sum which is unknown. Since it is impossible to know in real world problems the true sum, we could only provide the interval of doubt for the final desired result/number. The larger the interval of doubt is, the less reliable is the result.

In the case of floating-point numbers, the corresponding interval of doubt will be even larger. The floating-point number a = (.1134000, 0) would represent any real number x in the interval of doubt (of length 10^{-4}) .11335000 . . . ≤ x ≤ .11345000 . . . . The interval of doubt (of length 10^{-6}) for the floating-point number b = (.1134, −2) is .0011335000 . . . ≤ y ≤ .0011345000 . . . . The sum of intervals is .1145835000 . . . ≤ z ≤ .1146845000 . . . which has the length ($10^{-4} + 10^{-6}$). This is roughly 50 times larger than the interval of doubt for the sum in the fixed-point format.

Bibliography

ANSI/IEEE Standard 754-1985, Standard for Binary Floating Point Arithmetic

Alam, S.S.; Sen, S.K. (1996): *Computer and Computing with Fortran 77*, Oxford & IBH, New Delhi.

Ashenhurst, R.L.; Metropolis, N. (1959): Unnormalized floating-point arithmetic, *J. ACM*, 6, 415-28.

Ashenhurst, R.L.; Metropolis, N. (1965): Computers and computing, *AMM Slaught Memorial Papers*, 10, 47-59.

Barnett, D. (1987): A portable floating point environment, Unpublished manuscript.

Brown, W.S. (1981): A simple but realistic model of floating-point computation, *ACM Trans. Math. Software*, 7, 4, 445-480.

Chartres, B. A. (1966): Automatic controlled precision calculations, *J. ACM*, 13, 386-403.

Cody, W.J. et al. (1984): A proposed radix and word-length standard for floating point arithmetic, *IEEE Micro*, 4. 4, 86-100.

Cody W.J. (1988): Floating point standards — Theory and practice, In *Reliability in Computing: The Role of Interval Methods on Scientific Computing*; R. E. Moore, Ed., Academic Press, Boston.

Coonen, J. (1984): Contributions to a proposed standard for binary floating point arithmetic, Ph.D. dissertation, University of California, Berkeley.

Crandall, R.; Fagin, B. (1994): Discrete weighted transforms and large-integer arithmetic, *Math. Comp.*, 62, 305-324.

Dekker, T.J. (1971): A floating point technique for extending the available precision, *Numer. Math.*, 18, 3, 224-42.

Demmel, J. (1984): Underflow and the reliability of numerical software, *SIAM J. Sci. Stat. Comput.*, 5, 4, 887-919.

Dwyer, P.S. (1951): *Linear Computations*, Wiley, New York.

Farnum, C. (1988): Compluiler support for floating point computation, *Software Pract. Experi.*, 18, 7, 701-709.

Forsythe, G.E.; Moler, C.B. (1967): *Computer Solution of Linear Algebraic Systems*, Prentice-Hall, Englewood Cliffs, New Jersey.

Forsythe, G.E. (1970): Pitfalls in computation or why a math book isn't enough, *Amer. Math. Monthly,* 77, 931-56.

Gibb, A. (1961): Procedures for range arithmetic, , Algorithm 61, *Comm. ACM*,4, 319-20.

Goldberg, I. B. (1967): 27 bits are not enough for 8-digit accuracy, *Comm. ACM*, 10, 2, 105-06.

Goldberg, D. (1990): Computer Arithmetic. In *Computer Architecture: A Quantitative Approach*, D. Patterson and J.L. Hennessy, Eds., Morgan Kaufmann, Los Altos, California, Appendix A.

Goldberg, D. (1991): What every computer scientist should know about floating-point arithmetic, *ACM Computing Surveys*, 23, 1, 5-48.

Goldstein, M. (1963): Significance arithmetic on a digital computer, *Comm. ACM*, 6, 111-17.

Golub, G.H.; Van Loan, C.F. (1989): *Matrix Computations*, The John Hopkins University Press, Baltimore.

Gregory, R.T.; Krishnamurthy, E.V. (1984): *Methods and Applications of Error-free Computation*, Springer-Verlag, New York.

IEEE 1987. IEEE Standard 754-1985 for Binary Floating Point Arithmetic, IEEE Reprinted in SIGPLAN 22, 2, 9-25.

Kahan, W. (1972): A Survey of Error Analysis, In *Information Processing, 71* , North Holland, Amsterdam, vol. 2, 1214-1239.

Kahan, W. (1986): Calculating area and angle of a needle-like triangle, unpublished.

Kahan, W. (1987): Branch cuts for complex elementary functions. In the *State of the Art in Numerical Analysis*, M.J.D. Powell and A. Iserles, Eds., Oxford University Press, Chap. 7.

Kahan, W.; Coonen, T.J. (1982): The near orthogonality of syntax, semantics, and diagnostics in numerical programming environments. In *The Relationship between Numerical Computation and Programming Languages,* J. K. Reid, Ed., North-Holland, Amsterdam, 103-115.

Kahan, W.; LeBlanc, E. (1985): Anomaliesin the IBM acrith package. In *Proceedings of the 7th IEEE Symposium on Computer Arithmetic* (Urbana, Illinois), 322-331.

Kirchner, R.; Kulisch, U.W. (1987): Arithmetic for vector processors. In *Proceedings of the 8th IEEE Symposiumon Computer Arithmetic* (Italy), 256-69.

Knuth, D.E. (1969): *The Art of Computer Programming* (Vol. 2), Addison-Wesley, Reading, Massachusetts.

Knuth, D.E. (1981): *The Art of Computer Programming*, Vol. 2, 2nd ed. Addison-Wesley, Reading, Massachusetts.

Krishnamurthy, E.V. (1965): On a divide-and-correct method for variable precision division, *Comm. ACM*, 8, 179-81.

Krishnamurthy, E.V. (1970a): On optimal iterative schemes for high-speed division, *IEEE Trans. Computers*, C-20, 470-72.

Krishnamurthy, E.V. (1970b): A more efficient range-transformation algorithm for signed digit division, *Int. J. Control*, 12, 73-79.

Krishnamurthy, E.V. (1970c): Carry-borrow free sequential quotient generation with segmented signed digit operands, *Int. J. Control*, 12, 81-93.

Krishnamurthy, E.V. (1970d): On range transformation techniques for division, *IEEE Trans. Computers*, C-19, 157-60.

Krishnamurthy, E.V. (1971a): Complementary two-way algorithms for negative radix conversion, *IEEE Trans. Computers*, C-20, 543-50.

Krishnamurthy, E.V. (1971b): Economical iterative range transformation schemes for division, *IEEE Trans. Computers*, C-19, 179-81.

Krishnamurthy, E.V.; Nandi, S.K. (1967): On the normalization requirement of divisor in divide-and-correct methods, *Comm. ACM*, 10, 809-13.

Krishnamurthy, E.V.; Rao, T.M.; Subramanian, K. (1975a): Finite segment p-adic number systems with applications to exact computation, *Proc. Ind. Acad. Sci.* 81a, 58-79.

Krishnamurthy, E.V.; Rao, T.M.; Subramanian, K. (1975b): p-adic arithmetic procedures for exact numerical computation, *Proc. Ind. Acad. Sci.*, 82A, 165-75.

Krishnamurthy, E.V.; Sen, S.K. (2001): Numerical Algorithms: Computations in Science and Engineering, Appiliated East-West Press, New Delhi.

Kulisch, U.W.; Miranker, W.L. (1986):The arithmetic of the digital computer: a new approach, *SIAM Rev.*, 28, 1, 1-36.

Lakshmikantham, V.; Maulloo, A.K.; Sen, S.K.; Sivasundaram, S.(1997): Solving linear programming problems exactly, *Appl. Math. Comput.*, 81, 69-87.

Matula, D. W.; Kornerup, P. (1985): Finite precision rational arithmetic: slash number systems, *IEEE Trans. Computers*, C-34, 1, 3-18.

Metropolis, N.; Ashenhurst, R.L. (1963): Basic operations in an unnormalized arithmetic system, *IEEE Trans. Computers*, EC-12, 896-904.

Moore, R.E. (1961): Interval Analysis, Prentice-Hall, Englewood Cliffs, New Jersey.

Nandi, S.K.; Krishnamurthy, E.V. (1967): A simple technique for digital division, *Comm. ACM*, 10, 299-301.

Rajaraman, V.; Radhakrishnan, T. (1983): An Introduction to Digital Computer Design, 2nd ed., Prentice-Hall of India, New Delhi.

Rao, T.M. (1975): Finite-field Computational Techniques for Exact Solution of Numerical Problems, Ph.D. Dissertation, Department of Applied Mathematics, Indian Institute of Science, Bangalore.

Reisser, J.F.; Knuth, D.E. (1975): Evading the drift in floating point addition, *Inf. Process Lett.*, 3, 3, 84-87. 14, 111-112.

Rokne, J.; Lancaster, P. (1971): Complex interval arithmetic, *Comm. ACM,*

Sankar, P.V.; Chakrabarti, S.; Krishnamurthy, E.V. (1973a): Arithmetic algorithms in a negative base, *IEEE Trans. Computers*, C-22, 120-25.

Sankar, P.V.; Chakrabarti, S.; Krishnamurthy, E.V. (1973b): Deterministic division algorithm in a negative base, *IEEE Trans. Computers*, C-22, 125-28.

Sen, S.K. (2003): Error and computational complexity in engineering, in *Computational Mathematics, Modelling and Algorithms*, ed. J.C. Misra, Narosa Pub. House, New Delhi.

Sen, S.K.; Jayram, N.R. (1980): Exact computation of a matrix symmetrizer using p-adic arithmetic, *J. Indian Inst. Sci.*, 62A, 117-128.

Swartzlander, E.E.; Alexopoulos, G. (1975): The sign/logarithm number system, *IEEE Trans. Comput.* C-34, 12, 1238-42.

Venkaiah, V. Ch. (1987): Computation in Linear Algebra: A New Look at Residue Arithmetic, Ph.D. Dissertation, Department of Applied Mathematics, Indian Institute of Science, Bangalore.

Venkaiah, V. Ch.; Sen, S.K. (1987): A floating-point-like modular arithmetic for polynomials with application to rational matrix processors, *Adv. Modelling and Simulation*, 9, 1, 1-12.

Venkaiah, V. Ch.; Sen, S.K. (1988): Computing a matrix Symmetrizer exactly using multiple modulus residue arithmetic, *J. Comput. Appl. Math.*, 21, 27-40.

Venkaiah, V. Ch.; Sen, S.K. (1990): Error-free matrix symmetrizers and equivalent symmetric matrices, *Acta Applicande Mathematicae*, 21, 291-313.

Wijngaarden, A. van (1966): Numerical analysis as an independent science, *BIT*, 6, 66-81.

Wilkinson, J.H. (1963): *Rounding Errors in Algebraic Processes*, Prentice-Hall, Englewood Cliffs, New Jersey.

Wilkinson, J.H. (1965): *Algebraic Eigenvalue Problem*, Clarendon Press, Oxford.

Chapter 5

Error and Complexity in Numerical Methods

5.1 Introduction

5.1.1 Error and complexity: Brief review

When a quantity or a computation involving quantities is not exact, error creeps in. An error in a quantity is simply the difference between its exact and its approximate representations. Unless a quantity is discrete and is measured in terms of the number of items or pieces correctly (exactly), it is always in error since a measuring device can never measure a physical (real) quantity exactly. If the number of items/pieces is too large, say 10^6, then we may not represent this number exactly. The number of red blood cells 4.71 million per mm^3, for instance, is not measured exactly. Therefore, an error, although undesired, can never be gotten rid of. Further, since the exact quantity is never known, the absolutely true error (not the error-bound) is never known. What we know of is possibly a quantity Q of higher order accuracy and the quantity Q′ of lower order accuracy. The quantity of higher order accuracy should be such that the exact or, equivalently, absolutely true quantity lies in the interval $[-|Q - Q'|, +|Q - Q'|]$. We know this bound with 100% confidence unlike that in statistics, which can be known with, say 95% or 99% or even 99.9% confidence and never with 100% confidence. The foregoing interval should be as narrow (small) as possible to be as meaningful as possible. The 100% confidence in statistics usually implies too large or an infinite interval in which the exact quantity lies. Such an interval representing an error-bound is of no use in practice.

We have stressed here that error though unwanted remains an integral part of any real quantity and any computation involving real quantities. Any real quantity in nature is always errorfree but we never know its absolute correct (i.e., errorfree) value, in general. However, the knowledge of the error is necessary to get a logical feel of the quality of the quantity/computed result. In practice, too much of accuracy or, equivalently, too less of error in a quantity or in a computation is usually unnecessary. This is because such an accuracy will not, in general, make any difference in a real-world implementation. Thus the knowledge of error would save computing resources in terms of computing time and storage. In addition, it would establish how good the quantity or the computed result is. This saving is achieved due to avoidance of additional computation neeeded for more accuracy beyond certain meaningful/usable limit. We present, in the subsequent sections, the different kinds of errors and their computations associated with various algorithms implemented on a digital computer. We stress the fact that *anybody involved in scientific and engineering computations with data which are invariably erroneous should compute associated errors to satisfy oneself and others the goodness of all the computations done.*

It is also necessary to know how expensive the computation is. It is implicitly assumed that the computational problem is sufficiently large. The amount of computation that is used to measure the computational complexity (of an algorithm) needed by an algorithm (formal method) to solve the problem is an important parameter, besides however the error associated with it, to decide on the scope and quality of the algorithm. *We, in general, do not judge an algorithm based on its performance (computational complexity) in solving small problems.* ***In science and engineering computations, the two parameters, viz., the error and the computational (or time) complexity and sometimes even space complexity associated with an algorithm should be computed/known***. These will enable us to get a logical feel about how good the result is as well as how fast the algorithm is. We would also be able to compare two or more algorithms to solve the same type of problems. We integrate the complexity of an algorithm with the error appropriately in several places of this chapter. The ***space complexity***, i.e., the amount of storage/memory locations needed by the program/algorithm and the data, though important, will not be discussed here. However, the program size, in general, is independent of the size of the problem, i.e., data representing the problem. If we assume the size of the program negligible (a reasonable assumption), then the space complexity will be that of the data. For example, to multiply two $n \times n$ matrices, we need $O(2n^2)$ storage space for the data. No separate storage space would be used to store the product matrix.

Any quantity in the material universe (MU) is errorfree (exact). Any quantity that is produced through interaction/reaction of two or more quantities in the MU or natural processes is exact. ***Error is thus nonexistent in the MU and in all its (natural) activities/manifestations/processes***. However, we are not able to represent the real quantities of the MU exactly unless these quantities are discrete and are measured in terms of number of items.

Any quantity that is generated due to the environmental activities in the MU is also exact although we are not able to express or represent or even know this quantity exactly. The foregoing environmental activities correspond to some kind of computation.

The material universe (MU) has always a perfect order in it. All the laws of the MU are laws of nature (of matter) governing the order. We have unearthed some of the laws possibly not always exactly, but many are yet to be unearthed/understood by us. There is absolutely no inconsistency in this order of the MU. That is, ***the material universe is not only absolutely error-free and but also noncontradictory (consistent)***. The preceding statement is an *axiom.* All the events that happen in it follow the laws of nature and natural processes. These, of course, sometimes or even most time could be beyond our comprehension. In essence, we have never conclusively discovered/experienced violation of the laws and the processes. See also Section 8 of Chapter 1.

Human beings from the very dawn of civilization are inquisitive and have tried to get answer/solution to numerous queries/problems that have cropped up in their scientific/logical minds. Besides several motives behind this inquisitiveness, a dominant motive is how they can make best use of the MU for the good/benefit of mankind. Thus created by human beings from the MU is a physical problem (PP) — also called a physical model. A mathematical model (MM) — simulation/nonsimulation — is then derived from the PP by imposing assumptions. The MM is then translated into an appropriate algorithm (ALG), i.e., a method of solution and subsequently into a computer program (CP) – say, MATLAB, FORTRAN, C, C++, or JAVA program. The digital computer then takes this CP — call it CP_1 (usually a high-level program written in, say C or C++ or MATLAB or FORTRAN 90 language) — as the input and translates this program into the machine program — call this program CP_n (for a two-pass compiler, CP_2 could be an internal machine representation in reverse Polish notation and CP_3 the machine language program for the target machine, i.e., n here could be 3) — via one or more intermediate/internal machine representations — call these representations CP_2, CP_3, ... Finally the computation (COMP), i.e., the execution of machine program takes place and the results are produced. Errors ($\geq$ 0) are introduced in all the foregoing stages starting from the PP and ending in COMP (see Figure 1.4 in Section 9 of Chapter 1).

While the concerned problem of the MU is exact, the corresponding PP has the error due to (i) inexact representation of the physical quantities and (ii) assumptions that are needed to reduce complexity of the natural problem of the MU as well as to enable us to devise an algorithm more easily. Each one of the successive stages, viz., MM, ALG, CP_1, . . , CP_n (machine program), COMP injects error (or has the potential to inject error) into the system of problem-solving so that the RESULT (output) contains the cumulative error. This cumulative error is not necessarily always greater than the preceding errors since the true errors could be on the positive side or on the negative side. We, however, will never know whether an error is on the positive side or on the negative side. While computing the error bounds that should bracket the exact solution, we take errors as nonnegative, rather positive quantities.

The stages PP, MM, ALG, . . , CP_n are equivalent in the sense that each one is an ordered set of imperative sentences (instructions). The machine language (computer) program CP_n thus will have the largest number of instructions which the hardware machine (computer) understands, executes (COMP), and then produces RESULT See also Section 10 of Chapter 1.

5.1.2 Error due to instability in numerical computation

Error-free arithmetic, such as multiple-modulus residue arithmetic, p-adic arithmetic, rational arithmetic (practically not used because of *intermediate number growth*) could be employed only when the inputs are rational (ratio of two finite integers) and the number of operations in the algorithm used is finite (Gregory and Krishnamurthy 1984). For an *ill-conditioned* problem (a problem whose solution produced by using a finite-precision real/floating-point arithmetic has highly pronounced error) involving rational inputs, inexact arithmetic, viz., the real or the complex floating-point arithmetic produces highly erroneous results. This fact, usually known as the numerical instability, has been demonstrated by the following example (Sen 2002). The correct (exact) solution of the linear system

$$129838242x - 318037442y = 2, \quad 83739041x - 205117922y = 0$$

is $x = 102558961$, $y = 41869520.5$ while the computer outputs $x = 106018308.007132$, $y = 43281793.001783$ using Matlab (inv command, 15 digit accuracy). The superimposition of an error-free arithmetic on an algorithm is thus not only desirable but also often a must for many ill-conditioned problems. The only assumption for the exact computation, that is made is that the input rational data are exact although such an assumption is usually not valid for most real-world problems.

5.1.3 Error in the output of errorfree computation

Even in error-free implementation (which is possible when the algorithm/method involves only a finite number of arithmetic operations, viz., addition, subtraction, multiplication, and division), the inherent unavoidable error in the input quantities gets magnified in the output results although the computation is 100% exact. The authors are not aware of any practically useful study that has been made in this regard. However, this error could be studied using the ***interval arithmetic*** [Rokne 1971] although sometimes the interval in which the exact quantity lies becomes so large that it is not attractive or useful in practice. This study may be sometimes useful but significantly more expensive. An error estimate under specified (< 100%) confidence level is only possible using a polynomial-time probabilistic (randomized) algorithm. A deterministic algorithm (Sen and Mohanty 2003) to compute the error-bounds in an error-free computation is exponential and hence intractable i.e., solvable in Turing sense (Section 3.2.5 of Chapter 3) but prohibitive due to enormous computation time required.

In Section 5.2, we have mentioned different errors and their origin in quantities and in numerical computations and its importance. The complexity of algorithms including order of magnitude, hardness of a problem, and fastness of an algorithm is discussed in Section 5.3. Error and approximation in a computer including significance, floating-point arithmetic, different kinds of error and safeguards against them are presented in Section 5.4 while Section 5.5 comprises several algorithms with the related error and complexity.

5.2 Error in quantities and computations

We have already discussed what an error — relative, percentage, and absolute — is in Section 2.2 of Chapter 2, where we also have discussed about how to compute error since the exact quantity is never known. In numerical computation, ***it is the relative error that is almost always used while the absolute error is often not used.*** In Chapter 2, we have also noted that a ***measuring instrument is erroneous with a fixed order of error and this order varies from one measuring instrument to another***. Further, ***almost all measuring instruments will have an error (relative) of the order not less than 0.005 % (i.e., 0.5×10^{-4}***). This implies that it is not of much use in practice to produce/compute final numerical results with a relative error less than 0.5×10^{-4}. Thus most of the time for the final numerical (iterative) solution, we can introduce test (in a computer program) whether the relative error in a quantity falls below 0.5×10^{-4} or not. It is not necessary in practice for the physically used quantity to have its relative error less than 0.5×10^{-4} as it will not serve any purpose in any real world situation. However, in the intermediate steps, higher-order accuracy in computation would often be required so that the final result that will be used for actual engineering implementation is having error (i.e., the order of error) not less than 0.5×10^{-4}. To achieve a numerical computational error less than this quantity will have no other negative effect except the extra computing cost subject, however, to the precision (word-length) of the computer used.

5.2.1 Use/Importance of Error

Error, though absolutely unwanted, pervades in almost all our problems and can never be absolutely correctly known, nor can this be stopped from entering into our physical problems, algorithms, programs, and computations. Only its bounds are known usually. Can this error be of any use to us? The answer is *yes*. The result that we produce has to be validated for its quality. If it is not done then we would not know how good the result is. A logical validation is done by computing the error, i.e., the error-bound in the result. If the error-bound is reasonably narrow then the quality of the result is good. If it is not then the quality of the result may or may not be good. However, the logical error bound (reasonably narrow/small) may sometimes be difficult to be computed. One may say that the computed result may be verified/validated by the experimental result (in physics, chemistry, any engineering). True it is in an informal sense. However, there may not be a possibility of carrying out an experiment in an environment due to several constraints or an experimental result may not be available or *an experiment may be too expensive or too difficult or too hazardous*. There are numerous engineering problems which are solved/simulated without an experiment. *The error-bound of the result, when computed and reasonably narrow or sharp, is scientifically/logically an accepted means to establish the quality of the result*. If we do not have the knowledge of the error-bound of the result then we are in the dark and our faith in the result is shaky. While solving many physical problems numerically through solving the corresponding mathematical models which are partial differential equations, we may not easily compute the error-bounds. Yet we accept the result as valid possibly because we like them as these give 2 or 3-dimensional (graph) pattern that we expect. Though such an acceptance is not rigidly logical, it may serve the purpose in some sense. We cannot, however, be 100% sure that somebody in future may solve the same problem and obtain a different or a contradictory result.

5.3 Computational complexity

5.3.1 Order of Magnitude

The order of magnitude information provides us a good comparative feel about different quantities/functions (Sen 2002). It is, however, not absolutely necessary to associate error with order of magnitude. Consider the two functions $\phi(n)$ and $\psi(n)$. The symbols used in the language of comparing the rates of growth of function/computation (where $n \to \infty$ and $\exists$ = there exists) are as follows.

Symbol	Read as	Definition
o	is little oh of	$\phi(n) = o(\psi(n))$ if $\lim \phi(n)/\psi(n) = 0$
O	is big oh of	$\phi(n) = O(\psi(n))$ if $\exists C, n_0$ such that $\lvert\phi(n)\rvert < C\psi(n)$ (for all $n > n_0$)
~	is asymptotically[1] equal to	$\phi(n) \sim \psi(n)$ if $\phi(n)/\psi(n) = 1$
Ω	is omega of	negation of o

Examples

(i) $n^3 = o(n^5)$, $\sin(n) = o(n)$, $100 \log n = o(n^{.013})$.

(ii) $\cos(x) = O(x)$ or even more $\cos(x) = O(1)$, also $x^5 + 5x^3 + 43\cos(x) = O(x^9)$ or even more $x^5 + 5x^3 + 43\cos(x) = O(x^5)$ and $1/(1 + x^3) = O(1)$. In fact, we would usually use the 'big oh' notation O for stricter bounds such as $\cos(x) = O(1)$, $x^5 + 5x^3 + 43\cos(x) = O(x^5)$. Observe that 'o' gives more precise information than 'O' since we may write $1/(1 + x^2) = o(1)$ – this is typical of the relationship between O and o. Although quite often a 'O' result is sufficient for an application, we may sometimes need the more precise 'o' estimate. The "big-oh" notation allows us to replace the "≈" sign by "=", e.g., the Taylor series (infinite) for the function $f(x + h)$ can be written as

$$f(x+h) = f(x) + hf'(x) + (h^2/2!)f''(x) + \ldots + (h^n/n!)f^{(n)}(x) + \ldots \text{ to } \infty$$
$$\approx f(x) + hf'(x) + (h^2/2!)f''(x) + \ldots + (h^n/n!)f^{(n)}(x)$$

as well as

$$f(x+h) = f(x) + hf'(x) + (h^2/2!)f''(x) + O(h^3).$$

This is a ***one way equality***. We write $Cn^3 + Bn^2 + Dn = O(n^3)$. But we never write $O(n^3) = Cn^3 + Bn^2 + Dn$, where C, B, and D are constants.

(iii) $n^3 + n^2 + 1 \sim n^3$, $(2x + 1)^5 \sim 32x^5$, $(5x^5 + 40x^3 + 43)/(x^4 + 7x + 1) \sim 5x$, $7^x + 2\cos x - 4\log x \sim 7^x$, $\sin 1/x \sim 1/x$. In '~', the multiplicative constant has to be exactly correct.

---

[46] A series $s(x) = \Sigma\, a_k x^{-k}$, where Σ is the summation over $k = 0$ to ∞, is an asymptotic expansion or, simply an asymptote of a function $f(x)$ if $f(x) - s(x) = O(x^{-n})$ as $x \to \infty$ for every $n = 1, 2, \ldots$. We write $f(x) \sim s(x)$. The series itself may be convergent or divergent.

5.3.2 Hard/easy problem and slow/fast algorithm

Although error is not explicitly associated with the hardness of a problem or slowness of an algorithm, these are closely concerned terms which are preferably to be understood to appreciate the implication of error that could be inherent with the problem and the algorithm. Both the problem and the algorithm are an ordered set of imperative sentences (i.e., commands/instructions) with specified input data. The problem (e.g., the mathematical model) may be considered as a highest-level algorithm in some sense, say in the sense of imperative sentences used to write the mathematical problem).

Let L be the length of the input (data) in terms of the number of bits to represent the problem (or the algorithm). A problem will be called *easy* if it is guaranteed an answer that will be output after at most αL^n units of time/computation/work, where α is a positive real constant, and n is a finite positive real number. More generally, the problem (more formally, the class of problems) is easy if, whatever polynomial P we think of, there exists arbitrarily large values of L, and input data strings of L bits, it is guaranteed an answer that will be output after at most P(L) units of time or computation. Else, the problem is ***hard.***

If an answer is guaranteed only after e^L or $2^{P(L)}$ units of time or computation then the problem (rather, the class of problems) is ***hard***. A *performance guarantee* as mentioned above is sometimes termed as a *worst-case complexity estimate* or ***worst-case bound*** — this is the most commonly used estimate/bound. There is another kind of bound called ***average-case bound*** which states that if the performance is averaged over all possible input bit strings of L bits then the average amount of time or computation will give the average-case bound. See also Section 1.5 of Chapter 1.

Similarly, the slow and fast algorithms correspond to hard and easy (class of) problems, respectively. A slow algorithm is also termed as an ***exponential (not polynomial of finite degree) time*** algorithm while a fast algorithm is called a polynomial-time algorithm. Even among fast algorithms, some are faster than others. For example, an algorithm with computational complexity $O(2.5n^2)$ is faster that the one with complexity $O(0.333n^3)$. An algorithm whose worst-case complexity is e^L or $a^{P(L)}$ is termed ***exponential.*** It is desirable to work with algorithms where we can give a ***performance guarantee*** or, equivalently, complexity (computational/time) estimate that is at most a polynomial function of L. It is also desirable to work with algorithm where we can give a ***quality guarantee*** for the result/answer through an error (i.e., an error-bound which is rather sufficiently narrow to be meaningful in real-world situation). Thus, ***both the quality of performance of the algorithm (in terms of time/computational complexity of the algorithm) and the quality of the result (in terms of error bound of the result) should be***

known/attached along with the algorithm. These two information do specify the character of the algorithm and provide us its place/scope among other algorithms to solve the same class of problems.

However, a ***probabilistic algorithm***, in addition, should specify the probability of the correctness of the result. Often it is done by indicating the confidence level. Depending on the context, usually a 90% or a 95% or a 99% confidence level is used to indicate the (relative) amount of confidence that we can place on the computed results. An x% confidence level may also be termed as (100–x)% significance level. 99% confidence level, for instance, is exactly 1% significance level. Observe that the lower the significance level is, better (higher) is our confidence in the correctness of the result.

Consider the multiplication of two $n \times n$ matrices $A = [a_{ij}]$ and $B = [b_{ij}]$. We need n^3 multiplications and $n^2 \times (n - 1)$ additions to obtain the $n \times n$ product matrix $C = [c_{ij}] = [\Sigma a_{ik}b_{kj}]$, where the summation Σ is over $k = 1$ to n. Thus the computational complexity is n^3 multiplications + $n^2 (n-1)$ additions (without distinguishing between addition and subtraction) + other control operations/instructions that exist in the program (for matrix multiplications). Since n^3 is the most dominant term and is considered large, the computational complexity is written as $O(n^3)$. In this case the problem of matrix multiplication is *easy* and the conventional (straightforward) algorithm/method is *fast* or, equivalently *polynomial-time*.

A faster matrix multiplication is due to V. Strassen (Aho et al 1974). For the input 2×2 matrices A and B, we need, in the conventional multiplication, 8 multiplications and 4 additions to compute the product matrix $C = AB$. In Strassen method, the product matrix C has the elements, allowing $d = (a_{12} - a_{22}) \times (b_{21} + b_{22})$, $e = (a_{11} + a_{22}) \times (b_{11} + b_{22})$, $f = (a_{11} - a_{21}) \times (b_{11} + b_{12})$, $g = (a_{11} + a_{12}) \times b_{22}$, $h = a_{11} \times (b_{12} - b_{22})$, $p = a_{22} \times (b_{21} - b_{11})$, $q = (a_{21} + a_{22}) \times b_{11}$, $c_{11} = d + e - g + p$; $c_{12} = g + h$; $c_{21} = p + q$; $c_{22} = e - f + h - q$. This method needs 7 multiplications and 18 additions (instead of 4). If A, B are each $n \times n$ matrices and $n = 2^N$, where N is a positive integer, and A, B are each partitioned into four $2^{N-1} \times 2^{N-1}$ matrices $[A_{ij}]$, $[B_{ij}]$ and the product matrix C is similarly partitioned into four $2^{N-1} \times 2^{N-1}$ matrices $[C_{ij}]$ and use the foregoing 11 formulas replacing in them the lower case letters a, b, c by the upper case letters A, B, C, we will be able to appreciate the reduction of the number of multiplications when n is large. The computation of C, in Strassen method, needs $O(n^{\log 7}) = O(n^{2.8074})$ multiplications of numbers, where the base of the logarithm is 2. Observe that multiplications are more important than additions (counting also subtractions and not distinguishing between them). Because when the subprogram for matrix multiplication is called recursively, each multiply operation will be a multiplication of two big matrices while each addition will be an addition of two big matrices needing only $O(n^2)$ add operations (much cheaper). Several matrix multiplication algorithms have been developed during the last three decades (1970-

2000); the exponent has progressively come down to values less than 2.5. It appears that the exponent cannot be made 2 but possibly this can be made a value which is $2+\varepsilon$, where ε is *a small positive fraction*. What could be the minimum value of ε is not yet known and is thus an open problem. All the foregoing matrix multiplication algorithms are clearly fast algorithms – the one that is $O(n^{2.8074})$ is faster than the straightforward conventional one that is $O(n^3)$.

If each element of the matrices A and B is represented using 32 bits then the length of the input string is $L = 2 \times 32n^2 = 64n^2$ bits. Therefore, $n=(\sqrt{L})/8$. Hence the computational/time complexity (of the conventional matrix multiplication algorithm) measured in terms of the input bit-string size L is $O(\alpha L^{1.5})$, where $\alpha = 1/8^3$. Obviously this algorithm is ***fast***.

Now we consider, as an example, the problem of testing the primality of a given integer a. The straightforward conventional algorithm finds out whether a is divisible by $k = 2, 3, \ldots, \lfloor\sqrt{a}\rfloor$, where k should take ideally all the prime numbers in $[2, \lfloor\sqrt{a}\rfloor]$ one after the other instead of all the integers in $[2, \lfloor\sqrt{a}\rfloor]$. If all the answers are 'no' then a is a prime number. We need here $\lfloor\sqrt{a}\rfloor$ divisions (of an integer by another integer) if we divide a by each and every integer in $[2, \lfloor\sqrt{a}\rfloor]$. If the work is measured in terms of units of division, then we need $\lfloor\sqrt{a}\rfloor$ units of work. Here the length of the input bit-string $L = \log_2 a$, i.e., $a = 2^L$. The corresponding computational/time complexity of this straightforward algorithm, writing ln 2 for $\log_e 2$, is $O(2^{0.5L}) = O(e^{0.5L\ln 2})$ which is exponential (not polynomial). Hence this algorithm is ***slow*** and not desirable to be used although for small a this is good enough.

However, there are several probabilistic algorithms which are fast (polynomial-time) but the correctness of the output cannot be ascertained with 100% surety; this correctness can be said with a probability less than 1 and in several applications such a probabilistic result is useful. However, a polynomial time algorithm (Agrawal 2002) for testing the primality of a given integer has been devised (see Section 3.6 of Chapter 3).

Consider yet another problem — a linear program (LP) "*Compute the solution vector* $\boldsymbol{x}$ *that minimizes* $\boldsymbol{c}^t\boldsymbol{x}$ *subject to* $A\boldsymbol{x} = \boldsymbol{b}$, $\boldsymbol{x} \geq \boldsymbol{0}$ (null column vector), where $\boldsymbol{c}$ is an n numerically specified column vector, A is an $m \times n$ numerically specified matrix ($n \geq m$) of rank m, $\boldsymbol{b}$ is an m numerically specified column vector, and t denotes the transpose".

From the fundamental theorem[2] of linear programming, there are ${}^nC_m = n!/(m!(n-m)!)$ ways of selecting m of n columns (of A and of $\boldsymbol{x}$), and

---

[2] The fundamental theorem of linear programming is as follows. Consider

hence ${}^{n}C_{m}$ solutions of the linear system $Ax = b$, where $n \geq m$. One of these finite number of solutions will be the required solution of the LP provided the nonnegativity condition $x \geq 0$ is satisfied and there is a finite minimum value of the objective function. Let the LP be *Compute* $\mathbf{x}=[x_1 \;\; x_2 \;\; x_3 \;\; x_4]^t$ *that minimizes* $z = \mathbf{c}^t\mathbf{x} = [1 \;\; -2 \;\; 3 \;\; 1]\mathbf{x}$ *subject to* $A\mathbf{x} = \mathbf{b}, \mathbf{x} \geq \mathbf{0}$, where

$$A = \begin{bmatrix} 1 & 9 & 3 & 4 \\ -7 & 1 & -2 & 6 \end{bmatrix}, \quad \mathbf{b} = \begin{bmatrix} \mathbf{7} \\ \mathbf{0} \end{bmatrix}.$$

Here m=2, n=4, rank(A)=m=2. Hence there are ${}^{4}C_{2}=4!/(2!(4-2)!)=6$ ways of selecting 2 of 4 columns of A and of $\mathbf{x}$ and thus 6 solutions of the linear system $A\mathbf{x}=\mathbf{b}$:

$$\begin{bmatrix} 1 & 9 \\ -7 & 1 \end{bmatrix}\begin{bmatrix} x_1 \\ x_2 \end{bmatrix} = \begin{bmatrix} 7 \\ 0 \end{bmatrix}, \begin{bmatrix} 1 & 3 \\ -7 & -2 \end{bmatrix}\begin{bmatrix} x_1 \\ x_3 \end{bmatrix} = \begin{bmatrix} 7 \\ 0 \end{bmatrix}, \begin{bmatrix} 1 & 4 \\ -7 & 6 \end{bmatrix}\begin{bmatrix} x_1 \\ x_4 \end{bmatrix} = \begin{bmatrix} 7 \\ 0 \end{bmatrix},$$

$$\begin{bmatrix} 9 & 3 \\ 1 & -2 \end{bmatrix}\begin{bmatrix} x_2 \\ x_3 \end{bmatrix} = \begin{bmatrix} 7 \\ 0 \end{bmatrix}, \begin{bmatrix} 9 & 4 \\ 1 & 6 \end{bmatrix}\begin{bmatrix} x_2 \\ x_4 \end{bmatrix} = \begin{bmatrix} 7 \\ 0 \end{bmatrix}, \begin{bmatrix} 3 & 4 \\ -2 & 6 \end{bmatrix}\begin{bmatrix} x_3 \\ x_4 \end{bmatrix} = \begin{bmatrix} 7 \\ 0 \end{bmatrix},$$

$$[x_1 \;\; x_2]^t = [0.1094 \;\; 0.7656]^t, [x_1 \;\; x_3]^t = [-0.7368 \;\; 2.5789]^t,$$
$$[x_1 \;\; x_4]^t = [1.2353 \;\; 1.4412]^t, [x_2 \;\; x_3]^t = [\; 0.6667 \;\; 0.3333]^t,$$
$$[x_2 \;\; x_4]^t = [0.8400 \;\; -0.1400]^t, [x_3 \;\; x_4]^t = [\; 1.6154 \;\; 0.5385]^t.$$

In the first equation, x_1, x_2 are the ***basic variables*** while x_3, x_4 are the ***non-basic variables whose values are taken as zero*** in the original equation $A\mathbf{x} = \mathbf{b}$. In the second equation, x_1 is negative; while x_2, x_4 are the nonbasic ***variables*** whose values are taken as zero in the original equation $A\mathbf{x}=\mathbf{b}$. Since this solution does not satisfy the nonnegativity condition, we reject this solution. In the third equation, x_1, x_4 are basic variables and x_2, x_3 are nonbasic variables whose values are taken as zero. Thus there are four solutions, viz., the first, the third, the fourth, and the sixth solutions, each of which satisfies the nonnegativity condition. If we compute the objective function value $z = \mathbf{c}^t\mathbf{x}$ for each of the four values of the solution vector $\mathbf{x}$ then we obtain the value of z as −1.4218, −1.6471, 0, 0.5384, respectively. The minimum value

---

the LP Minimize $\mathbf{c}^t\mathbf{x}$ subject to $\mathbf{Ax} = \mathbf{b}, \mathbf{x} \geq \mathbf{0}$, where A is an $m \times k$ matrix ($k \geq m$) of rank m. If there is a (feasible) solution then there is a basic (feasible) solution, and if there is an optimal (feasible) solution then there is an optimal basic (feasible) solution. For proof, see Luenberger (1973).

of $z = -1.6471$ which corresponds to the third equation. Therefore, $\boldsymbol{x} = [x_1\ x_2\ x_3\ x_4]^t = [1.2353\quad 0\quad 0\quad 1.4412]^t$ is the required solution of the LP.

This algorithm is ***combinatorial*** (not polynomial-time) and thus is slow. Observe that the computational complexity to solve ${}^nC_m$ linear systems through the inversion of the square matrices, each of order m, is $O(m^3\times{}^nC_m)$, where ${}^nC_m = n!/((n-m)!\ m!)$, and $n! \approx (n/e)^n\sqrt{(2\pi n)}$ from the Sterling formula. If we solve the linear systems without inversion, say, using the Gauss reduction with partial pivoting then the complexity will be $O(m^3\times{}^nC_m/3)$ which is still exponential as it should be.

There was no polynomial-time algorithm for solving LPs till 1978. Since 1979, several polynomial-time algorithms for solving LPs have been developed. L. Khachian developed a polynomial-time algorithm (Khachian 1979), called Ëllipsoid algorithm, for LPs in integer model. N. Karmarkar designed a projective transformation based algorithm (Karmarkar 1984) which is polynomial-time, i.e., $O(n^{3.5})$ and is valid for real number models. These polynomial algorithms are fast while some are faster than the others. For solving small LPs, a slow algorithm may be more economical than the fast ones. Yet we would be interested in the fast ones and not in the slow ones since our main goal is to solve truly large problems. In fact, with the advent of high-performance computing devices, solving small problems is never a serious issue. ***The desired goal is to have a fast algorithm for truly large problems where slow algorithms will certainly be unacceptably expensive and thus useless.***

So far as the error in computation is concerned, it may not be more (or less) for slow (or fast) algorithm. The foregoing slow algorithm for the LPP is essentially solving the ${}^nC_m$ linear equations. The Gauss reduction method with partial pivoting (Krishnamurthy and Sen 2000) could do the job as a part of the slow algorithm and this could be implemented error-free provided the input data, viz., the elements of A, $\boldsymbol{b}$, and $\boldsymbol{c}$ accepted by the computer are considered exact. Even if it is not implemented error-free, the slow algorithm would almost always produce least error while a fast algorithm such as that due to Karmarkar [Karmarkar 1984] may not produce an error less than the least error.

5.4 What computer can represent

5.4.1 Computer representable numbers

Computers do not use the real number system. They use a simulation of it, called the floating-point number system. According to this system, a number is represented as an ordered pair of numbers in which the first number is a fraction or an integer called mantissa while the second one is an exponent. Sometimes this ordered pair may be represented the other way round. However, whatever way this number is represented in a computer, the same way (pattern) only is followed and is valid throughout. A floating-point number corresponds to a rational number. ***Further, only a very small fraction of rational numbers or, in other words, only a negligibly few rational numbers out of countably infinite rational numbers can be represented as floating-point numbers***. For example, 1/3 is a rational number but this may not be represented exactly as a floating-point number. Hence, these numbers, i.e., the computer representable numbers are countable and the total number of floating-point numbers is finite. These properties of the floating-point numbers are unlike those of real numbers ***(a real number is the totality of rational and irrational numbers)*** which are neither countable nor finite. For example, $\sqrt{17}$ is a real number which cannot be exactly represented as a floating-point number or even as a finite rational number (i.e., as the ratio of two finite integers) or, equivalently, a finite decimal number. Observe that the rational numbers are countable but infinite. Thus the floating-point number representation introduces an error called a rounding-off error.

5.4.2 Sources of error and safeguard against it

We have seen in Section 1.10 of Chapter 1 that error is introduced in all the stages starting from the physical model up to the computation including the intermediate steps of computation. The goal, therefore, will be to take enough precaution to eliminate the error or minimize it at every stage.

To ensure the ***reliability*** (implying nondominance of the error) of a mathematical model, one may (i) check the result with the real test problem – ***a test problem is one whose result is already known*** and (ii) examine the model in simple cases as well as in extreme cases.

To study the reliability of the algorithm, (i) check the result against those obtained independent of the algorithm, (ii) examine the algorithm in simple cases as well as in extreme cases, and (iii) compare the algorithms when there are two or more of them.

To ensure the ***stability*** of the algorithm, i.e., to ensure the accuracy in computation (i) do the computation with different amounts of precision (e.g., single, double, triple, . . . precision) in the arithmetic and (ii) solve several different problems whose initial data (input) are only slightly different. To discover the effect of rounding error in computation, solve, in addition, the same problem with arithmetic performed in different sequence. A simple way of achieving the different sequence may be to use two different compilers for the translation into the machine language. For an iterative algorithm, if an estimate gets closer to the answer then it will be a test for convergence of the algorithm.

The equality of two floating-point numbers cannot be easily established. This fact brings in a serious restriction in numerical algorithms. The equality $x_{i+1} = x_i$, where x_i is an iterate, *cannot be* tested. Thus, in iterative procedures involving infinite algorithms, we can only test whether $|x_{i+1}-x_i| \le \varepsilon|x_{i+1}|$, where ε is a suitable positive real number, say. 0.5×10^{-4} and may be called a ***numerical zero***. Note that we are testing whether the relative error $|x_{i+1}-x_i|/|x_{i+1}| \le \varepsilon$. An infinite (convergent) algorithm is one that generates a sequence x_i, where x_i will be usually closer to true value x (unknown) for increasing i . The Newton method (presented later) to compute a root of the equation f(x) = 0 is infinite.

5.4.3 Error in series approximation

A series is the sum of terms while a sequence, on the other hand, is the collection of terms. The sum of the terms in the sequence will be called a series. For example, $1, x, x^2/2!, x^3/3!, \ldots, x^n/n!, \ldots$, to ∞ is a sequence (infinite) while $1 + x + x^2/2! + x^3/3! + \ldots + x^n/n! + \ldots$ to ∞ is a series (infinite or power). If there is a finite number of terms in a sequence (or in a series) then the sequence (or the series) is finite. The term 1 in the sequence (or the series) is the 0-th term and the term $x_n/n!$ is the n-th term of the sequence (or the series). One may, however, call 1 as the first term and $x^n/n!$ as the (n +1)st term and proceed accordingly for a computation.

The series computation involves the addition of terms. This ***addition is not usually carried out by explicitly computing the value of each term and then adding them up***. It is carried out by expressing the (n+1)st term in terms of the n-th term and adding the (n + 1)st term to the already computed sum up to the n-th term. In the foregoing series, the n-th term is $t_n=x^n/n!$ and the (n+1)st term is $t_{n+1} = x^{n+1}/(n+1)!$. Hence the scheme for computing the value of the series $s = 1 + x + x^2/2! + x^3/3! + \ldots + x^n/n! + \ldots$ to ∝ is

$$s_0 = t_0 = 1,\ x = \text{a specified number (real or complex)},$$
$$t_{n+1} = t_n x/(n + 1) \text{ and } s_{n+1} = s_n + t_{n+1} \quad n = 0, 1, 2, \ldots, \text{ till } |t_{n+1}|/|s_{n+1}| \le 5 \times 10^{-5}$$

The value of s_{n+1} after the execution of the foregoing scheme is the required value of the series correct up to 4 significant digits. If we desire the value of the series correct up to 4 decimal places then we replace, in the foregoing scheme, $|t_{n+1}|/|s_{n+1}|$ by $|t_{n+1}|$. Observe that $|t_{n+1}|/|s_{n+1}|$ is the relative error while $|t_{n+1}|$ is the absolute error introduced due to the truncation of the infinite series after (n + 1)st term. Further, ***in numerical computation, we should almost always compute the accuracy in terms of significant digits and not* in *terms of decimal digits*** as explained earlier. In the foregoing computation we have assumed sufficiently large precision (word-length) of the computer used so that the rounding error is too small to affect the accuracy up to 4-th significant digit.

For the purpose of a computer implementation, we omit the subscripts to save storage space and write the ***computational scheme*** as

$$s = t = 1,\ x = \text{a specified number (real or complex)},$$
$$t = tx/(n+1) \text{ and } s = s + t \quad n = 0, 1, 2, \ldots, \text{ till } |t|/|s| \le 5 \times 10^{-5}$$

Here '=' implies 'is replaced by'. '=' is not mathematical equal to.

5.4,4 Speed of convergence

Convergence of an infinite (also called power) series could be fast or slow. Some diverges beyond certain range of values of the parameter x when the series is the function of x while some others are only conditionally convergent. The foregoing series is e^x and is fast convergent. ***To get an accuracy of 4 significant digits, if we do not require more than 5 or 6 terms of the infinite series for a specified range of values of x then the series is excellent. If, on the other hand, we need more than, say, 20 terms for a specified range of values of x then the series form of the function is not, in general, very desirable for numerical computation***. The computation of $\log_e(1 + x)$ by the series $x - x^2/2 + x^3/3 - x^4/4 + x^5/5 - \ldots$, to ∞, ($|x| \le 1$ and $x \ne -1$) is clearly undesirable for values close to 1 since it takes too many terms and hence too much of computation (introducing unacceptable error). For example, the number of terms required to obtain an accuracy of 4 decimal places is given by the inequality $|x^n/n| \le 5 \times 10^{-5}$, where the first term (not the zeroth term as is considered in the case of the series for e^x) is taken as x. When x = 0.98, the number of terms n is 223 (for 4 decimal digit accuracy). If x = 0.99, then n will be 392. These numbers of terms imply that too many terms of the series are needed when $\log_e(1 + x)$ is evaluated for values of x close to 1. One should, therefore, not use this series to evaluate the foregoing log function for x near 1. Either one should employ a faster convergent series for this function or use some other procedure. We have assumed that the precision of the computer is sufficiently large so that the rounding error is small compared to the truncation error (i.e., the error introduced due to the truncation of all the (infinite) terms after the first 223 terms or 392 terms depending on the value of x). The number of terms needed to get an accuracy of 4 significant digits is given by the inequality $|x^n/(n \times \log_e (1 + x))| \le 5 \times 10^{-5}$. When x = 0.96, the number of terms n is 133 which gives us an accuracy of 4 significant digits. If x = 0.99 then n = 392 for 4 decimal digit accuracy (as shown above) and n = 422 for 4 significant digit accuracy. We have taken $\log_e(1.96) = 0.6729$ and $\log_e(1.99) = 0.6881$. Observe that the value of $\log_e (1 + x) = \log_e (1.96)$ for x = 0.96 and that of $\log_e (1 + x) = \log_e (1.99)$ for x = 0.99 are not computable through the aforementioned series . Each of these values should be replaced by the current sum of the terms, which goes on changing with the addition of each new term (expressed in terms of the previous term).

For an infinite convergent series which is a function of x, to get an accuracy of 4 significant digits, the general scheme can be written as

$$t_0 = k \text{ (some specified numerical value)}, \quad s_0 = t_0,$$
$$t_{n+1} = t_n f(x) \text{ and } s_{n+1} = s_n + t_{n+1}, \quad n = 0, 1, 2, \ldots, \text{ till } |t_{n+1}|/|s_{n+1}| \le 5 \times 10^{-5}$$

where f(x) is found from the given series. For the series for e^x, $f(x)=x/(n+1)$.

5.5 Algorithms and related errors

5.5.1 Error in fixed-point iteration algorithms for equations

Most of the algorithms to compute a root of a nonlinear equation or, equivalently to solve a nonlinear equation are based on *fixed-point iterations.* In these algorithms, the given equation f(x) = 0 is rewritten in the form x = g(x). For example, if $f(x) = 3x^7 - 2x^2 + 1 = 0$ then we may take $g(x) = 3x^7 - 2x^2 + x + 1$ or $g(x) = (3x^7 + 1)/(2x)$ or $g(x) = x - f(x)/f'(x)$, where $f'(x) = df/dx = 21x^6 - 4x$ (*as in* ***Newton method***), or $g(x) = [(2x^2 - 1)/3]^{1/7}$ and so on. x = g(x) is then solved by the following successive substitution procedure.

$x_0 = \alpha$ (an appropriately chosen real (complex) number for real (real or complex) roots)
$x_{i+1} = g(x_i)$ i = 0, 1, 2, 3, . . ., till $|x_{i+1} - x_i|/|x_{i+1}| \le 5 \times 10^{-5}$ or i = 20

The choice of g(x) is important from the speed of convergence point of view. For some g(x), the iteration scheme converges fast while for others its convergence will be slow or it may not even converge, i.e., it may diverge or oscillate.

However, the iteration scheme will be meaningful only when it converges or sometimes it oscillates (for, say multiple roots) with a small amplitude. For a scheme that converges based on the choice of g(x) and the initial approximation for x, after a sufficient number of iterations, $g(x_i)$ remains fixed or converges at a value $x_i = p$ i.e., $p = g(p)$. Hence the name fixed-point iteration.

If a fixed-point iteration scheme needs too many iterations then it is inefficient. If the scheme takes 5 or 6 iterations to produce an accuracy of 4 significant digits then it is efficient and is considered good. However, in order to avoid too many iterations which would occur for slowly converging iteration schemes, we may include one additional stopping condition, viz., $i = 20$. This implies that if any fixed-point iteration scheme takes more than 20 iterations then we should relook into the scheme and attempt to modify $g(x)$ for faster convergence. If the choice of $g(x)$ is not good enough then the scheme may diverge.

If the (chosen) initial approximation x_0 is not reasonably close to the root/solution then also the scheme may diverge. ***If $f(x)$ is a polynomial, say, $f(x) = 3x^6 - 7x^5 + x - 20$ then for any finite value of x_0, the Newton scheme will converge.*** To obtain a complex root, x_0 should be chosen as a complex number and complex arithmetic should be used in the scheme. As an illustration, let us consider the equation $x^2 - 36 = 0$ and use the Newton (fixed-point iteration) scheme. Here $g(x) = x - f(x)/f'(x) = 0.5x + 18/x$. Hence the Newton (also known as the Newton-Raphson) scheme to solve a nonlinear equation $f(x) = 0$ in one variable x can be written as

$x_0 = \alpha = 36$ (chosen)

$x_{i+1} = x_i - f(x_i)/f'(x_i)$, $i = 0, 1, 2, \ldots$, till $|x_{i+1} - x_i|/|x_{i+1}| \le 5 \times 10^{-5}$.

The order of convergence (Krishnamurthy and Sen 2001) of this method is 2. Roughly speaking. ***this order implies that if a root of the equation $f(x)=0$ is found to be correct up to k ($k \ge 1$) digits in i-th iteration then the root should be correct up to about 2k digits in the (i+1)st iteration (assuming sufficiently large precision of the computer)***. The computation will proceed as follows. $x_0 = 36$ (not a good choice for the square-root of 36).

- $i = 0$, $x_1 = x_0 - f(x_0)/f'(x_0) = 0.5x_0 + 18/x_0 = 18.5$. Since $e_1 = |x_1 - x_0|/|x_1| = .9456 > 5 \times 10^{-5}$; go to the next step.
- $i = 1$, $x_2 = x_1 - f(x_1)/f'(x_1) = 0.5x_1 + 18/x_1 = 10.22297297297297$. Since $e_2 = |x_2 - x_1|/|x_2| = .8096 > 5 \times 10^{-5}$, we go to the next step. Thus, we obtain $x_3 = 6.87222673764313$, $e_3 = .4876$; $x_4 = 6.05535174484948$, $e_4 = .1349$; $x_5 = 6.00025298411942$, $e_5 = .0092$; $x_6 = 6.00000000533319$, $e_6 = 4.2167 \times 10^{-5} \le 5 \times 10^{-5}$ and stop. Hence a root of the equation $x^2 - 36 = 0$ is 6.00000000533319 which is correct at least up to 4 significant digits. The other root can also be found out by deflating the polynomial $x^2 - 36$, i.e., by dividing the polynomial by $x - 6.00000000533319$ (and, if necessary, by applying the Newton scheme once again on the deflated polynomial).

If there are ***repeated roots*** (e.g., the roots of the equation $x^3 - 6x^2 + 12x - 8 = 0$), then the Newton method will ***oscillate*** around the repeated root (root 2) without converging to the root (root 2). These is because of the fact that both f(x) and f′(x) become zero at x = 2, f(x) tends to 0 faster than f′(x), and the computer is finite-precision. In such a situation, one may use the deflated Newton method (Krishnamurthy and Sen 2001). To compute a complex zero of a real or a complex polynomial or a transcendental function (i.e., a polynomial of degree ∞, e.g., cos x, e^x, a function involving a combination of these functions) using the Newton scheme, we have to take as an initial approximation x_0 a complex value and use complex arithmetic. Using a real initial approximation and the real arithmetic, we will never get the required complex (including imaginary) root.

For a real root, we may use here the successive bisection method (Krishnamurthy and Sen 2001) instead of a fixed-point iteration method $x_{i+1} = \phi(x_i)$ by writing the equation f(x) = 0 as $x = \phi(x)$ which is always possible. ***There are numerous possible choices of $\phi(x)$.*** For example, we may choose $\phi(x)$ as x + f(x) or $\phi(x)$ as x − (f(x)/f′(x)) where f′(x) is the first derivative of f(x) with respect to x here. However, ***not all choices of $\phi(x)$ may converge*** for a specified initial approximation x_0. Also, not all convergent choices converge at the same speed. Some take more iterations while others take less.

We now construct quadratic polynomial $p(x) = x^2 + 19992.1000000001x - 78984.4000000005$ by imposing $p(x) = 10^3$ at x = 4 and $p(x) = -10^3$ at x = 3.9 (Sen 2002). This polynomial is ***highly sensitive or, equivalently, unstable***. Evidently, there is a real root of the equation p(x) = 0 for x in (3.9, 4) since the left-hand side of the equation is a finite degree (here degree 2) real polynomial and hence continuous and cut the x-axis for a value of x in (3.9, 4). The following table provides the values of x and the corresponding p(x).

x	3.9	4	3.975	3.9625	3.950000122	3.950000123
p(x)	-10^3	10^3	500	250	-6×10^{-5}	-4×10^{-5}

The computations have been performed with 16 significant digit floating-point arithmetic. Just by substituting the value of x in p(x), one might get, instead of 0, a large value, ±200 say, and might conclude that the computed root is wrong. Such a conclusion is inevitably incorrect as it is clear from the foregoing example. For a real-world problem, often the value of x correct up to 2 or 3 decimal places is good enough. Here, ***just by observing the change of sign in the value of p(x) for a relatively small change in the value of the root x, we should accept the root, although the value of p(x) is large***.

Besides the *stiff (i.e., violently) fluctuating polynomials* (i.e., the polynomials each of whose values differ too much for a small change in the value of x), there are ill-conditioned (with respect to computing zeros) polynomials having ***zero clusters*** (i.e., the polynomials each having closely spaced zeros).

For example, the polynomial p(x) that has 3.000, 3.001, 3.002, 3.003, 3.004, 3.005 as zeros, i.e., the polynomial equation p(x) = 0 that has 3.000, 3.001, 3.002, 3.003, 3.004, 3.005 as roots (constituting what is called a ***root-cluster***) is an ill-conditioned polynomial (with respect to computing roots) (Sen 2002). Computation of the roots in a root-cluster with reasonable/acceptable accuracy is difficult for any root-finding method. Observe that the exact roots in a root-cluster are not known a priori.

5.5.2 Interpolation

We may view interpolation as reading between lines in a table. There are two ways of defining a function y = f(x) – (i) analytically, e.g., $f(x) = 2x^2 - 5x.\cos x + 7$ and (ii) by a table (x_i, y_i) i = 0(1)n, where $x_0 < x_1 < x_2 < \ldots, < x_n$.

If the function f(x) is known analytically then, for a specified value of x, f(x) is readily computable; ***no interpolation is required***. Given a value of f, to compute the corresponding x is the inverse (not interpolation) problem and may not have a unique answer. The later inverse problem is essentially an equation-solving problem and clearly a tougher one.

If the function y = f(x) is given by the set of pair of values (x_i, y_i) i = 0(1)n or, equivalently, by the (rowwise) table

x	x_0	x_1	$x_2 \ldots x_{n-1}$	x_n
y	y_0	y_1	$y_2 \ldots y_{n-1}$	y_n

where x_i are in increasing order then to find the value of y for $x = \alpha$ in (x_k, x_{k+1}), where $k+1 \le n$, is the ***direct interpolation*** problem. The foregoing rowwise table can also be written as the columnwise table. We have written the table rowwise just to save space. To find the value of x for $y = \beta$ in (y_k, y_{k+1}) is the ***inverse interpolation*** problem. To find the value of y for $x = \alpha > x_n$ is the ***direct extrapolation*** problem while to find the value x for $y = \beta > y_n$ is the ***inverse extrapolation*** problem. Both the direct and the inverse interpolation problems are essentially identical – only the role of x and that of y need to be interchanged. So is the case with extrapolation problems. However, all these four problems are generally termed as simply ***interpolation*** problems.

To compute the error involved in the interpolation, consider the Lagrange interpolation formula. Given the table (x_i, y_i) i = 0(1)n, where x_i may or may not be equally spaced, get an n-th degree polynomial y(x) that passes through all the n+1 points (x_i, y_i). This polynomial is an approximation of the function f(x), which coincides with the polynomial at (x_i, y_i) i = 0(1)n. The required n-th degree polynomial (also known as the Lagrange interpolation polynomial) is

$$y(x) = \sum_{k=0}^{n} y_k P_k(x)/P_k(x_k), \quad \text{where } P_k(x) = \prod_{\substack{i=0\\ i\neq k}}^{n} (x - x_i) \quad k = 0(1)n$$

which is known as the ***Lagrange interpolation formula*** (Sen 2002; Krishnamurthy and Sen 2001). A relative error-bound in the formula is given by

$$E_r(x) = [|x_n - x_0|^{n+1} \max |f^{n+1}(\xi)|/(n+1)!]/y(x), \quad x_0 \le \xi \le x_n$$

where $f^{n+1}(x)$ is the (n+1)st derivative of the function f(x) with respect to x, ξ is a value in $[x_0, x_n]$ so that this derivative will be maximum in magnitude. ξ is not readily known; nor do we need to know ξ. All that we have to know is the largest value (in magnitude) of the (n+1)st derivative of f(x) in the interval $[x_0, x_n]$. Strictly speaking, the function f(x) is often not analytically known. If, for example, $f(x) = 2\cos^2(x) - 10$ then f(x) is considered analytically known. ***If f(x) is analytically known then there is usually no need to do interpolation.*** One can directly evaluate f(x) for the given value of x. Hence the ***foregoing error formula is not of much use in practice,*** particularly in the age of extensive availability of computing devices.

The function f(x) is known only in the form of the table (x_i, y_i) i = 0(1)n, where n could be large, say, 30. We are ***certainly not*** going to compute the 30-th degree Lagrange interpolation polynomial y(x) — that passes through all the 31 points correctly — to represent f(x). This is because we do not anticipate a violent fluctuation of the function f(x) between two consecutive points (x_k, y_k) and (x_{k+1}, y_{k+1}) for some k in [0, n−1]. On the other hand, we anticipate a smooth curve between two consecutive points. ***Almost always we use either first degree (linear) or second degree (quadratic) or third degree (cubic) Lagrange interpolation polynomial*** y(x) and not beyond although the table may have a large number of points (i.e., n is large, say, 30 or more). If we wish to use *linear (polynomial of degree 1) interpolation* to find y(x) for x in $[x_k, x_{k+1}]$ then we take only the two points (x_k, y_k) and (x_{k+1}, y_{k+1}), where k could be considered 0 and k+1 could be considered n (i.e., n=1) in the Lagrange interpolation formula. Thus, the ***Lagrange linear interpolation formula along with the relative error*** can be written as (Sen 2002)

$$y(x) = y_0(x-x_1)/(x_0-x_1) + y_1(x-x_0)/(x_1-x_0)$$
$$E_r(x) = [|x_1-x_0|^2 \max|f^2(\xi)|/2!|]/y(x), \; x_0\le\xi\le x_1$$

and the ***Lagrange quadratic interpolation formula with the relative error*** is (Sen 2002)

$$y(x) = y_0(x-x_1)(x-x_2)/((x_0-x_1)(x_0-x_2))+y_1(x-x_0)(x-x_2)/((x_1-x_0)(x_1-x_2))$$
$$+ y_2(x-x_0)(x-x_1)/((x_2-x_0)(x_2-x_1))$$
$$E_r(x) = [|x_2-x_0|^3 \max|f^3(\xi)/3!|]/y(x), \quad x_0 \le \xi \le x_2$$

where x is any value in $[x_0, x_1]$ or in $[x_1, x_2]$ or, equivalently, in $[x_0, x_2]$. If we omit y(x) or, equivalently, set y(x) = 1 in the foregoing two $E_r(x)$'s, we obtain the absolute errors. Similarly, we may write the Lagrange cubic interpolation formula along with the relative and absolute errors. Consider f(x)=cos x given as the table

x_0 (arrow to 0.1) $\qquad$ x_6 (arrow to 0.7)

x	0.1	0.2	0.3	0.4	0.5	0.6	0.7
y	0.9950	0.9801	0.9553	0.9211	0.8776	0.8253	0.7648

y_0 (arrow to 0.9950) $\qquad$ y_6 (arrow to 0.7648)

Use the Lagrange linear and quadratic interpolation formulas to compute y at x = 0.52 and the corresponding relative error. The solution is as follows.

Lagrange linear interpolation formula Here we consider x_0 = 0.5, y_0 = 0.8776, x_1=.6, y_1=0.8253, and x =0.52. Substituting these values in the linear interpolation formula and in the corresponding relative error formula, we get y(0.52) = 0.8776(0.52 − 0.6)/(0.5 − 0.6) + 0.8253(0.52 − 0.5)/(0.6 − 0.5) = 0.8671, $E_r(0.52) = [|0.6-0.5|^2 \times 0.8776/2!]/y(0.52) = 0.0051$. Observe that $|f^2(\xi)| = |d^2 (\cos x)/dx^2|_{x=\xi} = |-\cos x|_{x=\xi} = |\cos x|_{x=\xi}$ has the maximum value 0.8776 in the interval x_0 = 0.5 and x_1 = 0.6. The result, viz., y(0.52) = 0.8671 for cos (0.52) is correct up to 2 significant digits. The actual value of cos (0.52) up to 4 digits is 0.8678. Similarly, we may use quadratic as well as cubic interpolation formulas for better results.

Before the advent of the wide spread computer usage we were using the printed table (Abramowitz and Stegun 1965) for a function (special functions such as Bessel functions, Legendre functions, transcendental functions such as sine function, logarithmic functions) and using the values including the interpolated ones for further computations in complex mathematical expressions. Thus in such cases interpolations were useful although the functions are known analytically. However, such an interpolation for an analytically known function is often not needed as the computer provides the result readily, say, through the MATLAB software from the analytical expression. The current computational scenario thus makes obsolete numerous mathematical tables which were developed over decades and that did run over thousands of pages. However, tf a mathematical expression involves significant computation (more computation than interpolation) then interpolation might be resorted to if the function is known through a table of values.

In the real world situation we get values of x and corresponding y through certain measurements and we do not know the analytical form of the function for these values. Hence computing an error-bound is not possible using the foregoing formulas for E_r. However, what we can do is that ***we compute the interpolation result y_b using a lower order (linear, say) interpolation formula and the interpolation result y_g using a higher order (quadratic, say) interpolation formula to obtain the relative error as well as the absolute error***. The relative error is $|y_g - y_b|/|y_g|$ while the absolute error is $|y_g - y_b|$; ***the need to know the function analytically is thus obviated.*** Let $y_g = 0.4967$ and $y_b = 0.4963$. Hence the relative error in quadratic interpolation is $0.0004/0.4967 = 0.0008$ while the absolute error is .0004.

If we have to compute an x given a y (***inverse interpolation***) then the foregoing Lagrange interpolation formulae can be readily used just by interchanging the role of x and that of y. The ***extrapolation*** can also be carried out by these formulas considering the ***nearest*** *(x, y)* ***points.*** Consider the *physical problem*: For a particular feed, a chicken grows as in the following table. What will be its weight at the age of 3.3 months (Sen 2002)?

Age (month) x	1.5	2.0	2.5	3.0
Weight (kg) y	0.80	1.10	1.50	1.85

This is an extrapolation problem. If the ***Lagrange quadratic interpolation formula***

$$y(x) = [(x - x_1)(x - x_2)y_0/((x_0 - x_1)(x_0 - x_2))]+[(x - x_0)(x - x_2)y_1/((x_1 - x_0)(x_1 - x_2))] + [(x - x_0)(x - x_1)y_2/((x_2 - x_0)(x_2 - x_1))],$$

where $x_0 = 2.0$, $x_1 = 2.5$, $x_2 = 3.0$, $y_0 = 1.10$, $y_1 = 1.50$, $y_2 = 1.85$, $x = 3.3$, we obtain $y_q = y(3.3) = 2.0363$ kg.

For the relative error in y(3.3), we use the ***Lagrange linear interpolation formula***

$$y(x) = [(x - x_1)y_0/(x_0 - x_1)] + [(x - x_0)y_1/(x_1 - x_0)],$$

where $x_0 = 2.5$, $x_1 = 3.0$, $y_0 = 1.5$, $y_1 = 1.85$, $x = 3.3$. We obtain $y_l = y(3.3) = 2.0600$ kg. Hence the relative error in the linear interpolation (not the quadratic interpolation) is $|y_q - y_l|/|y_q| = 0.0116$.

When x = 3.2 months, the foregoing quadratic interpolation gives us y(3.2) = 1.9760 kg and the foregoing linear interpolation produces y(3.2) = 1.9900 kg. If x = 3.1 months, then the corresponding weights will be 1.9140 kg and 1.9200 kg, respectively.

Observe that ***for interpolation, we use the closest/nearest data points. Also note that the relative error in quadratic interpolation will be small compared to that in linear interpolation.*** The farther the data points are, the lesser is the effect of these points on interpolation. ***Almost always, linear or quadratic or at most cubic interpolation is used in practice***. To obtain the relative error in quadratic interpolation, we have to at least use cubic interpolation to know the higher order accurate y.

Interpolation in a table (x_i, y_i) $i = 0, 1, 2, \ldots, n$, (where n is large, say, 1000) having n+1 data points should never be done with the n-th (say, 1000th) degree polynomial. Although this polynomial passes through all the n + 1 points (say, 1001 points) exactly. In between two consecutive points, the curve represented by the n-th degree polynomial may have ups and downs which are undesired. ***Very violently fluctuating functions are ill-conditioned with respect to interpolation*** that, of course, includes extrapolation. One should take special care for such functions and should invariably compute a reasonably sharp error-bound to establish the quality of the result. To obtain a smooth curve through a set of points, cubic ***spline interpolation*** (Krishnamurthy and Sen 2001), fitting between each pair of points a cubic polynomial called a cubic spline, is used.

5.5.3 Linear system solver

Let $A\mathbf{x} = \mathbf{b}$ be the given system of linear equations, where $A = [a_{ij}]$ is an m × n numerically known matrix, $\mathbf{b} = [b_j]$ is a numerically known m-column vector. The problem is to find a value of the solution vector $\mathbf{x}$ and the relative error in $\mathbf{x}$ provided the system is consistent (i.e., $A\mathbf{x} = \mathbf{b}$ has one or infinite solutions). Let A^+ be the minimum norm least squares inverse which is also known as the Moore-Penrose inverse or the pseudo-inverse or the p-inverse (Lakshmikantham et al. 1996) of the matrix A. Then ***the general solution of*** $A\mathbf{x} = \mathbf{b}$ can be written as

$$\mathbf{x} = A^+\mathbf{b} + (I - A^+A)\mathbf{z},$$

where I = the n × n unit matrix and $\mathbf{z}$ is an arbitrary column vector of dimension n. This general solution will be a true solution if $A\mathbf{x} = \mathbf{b}$ is consistent else this will be the minimum norm least squares solution when $\mathbf{z}$ is taken $\mathbf{0}$ (null column vector). If the matrix A is nonsingular (and, of course, square), then the general solution will produce the unique solution $A^{-1}\mathbf{b}$.

Yet another problem is to compute the p-inverse A^+ and the relative error in it. ***The p-inverse A^+ will be the true inverse A^{-1} if the matrix A is non-singular, i.e., A is square and its determinant is not 0.*** The matrix $X = A^+$ satisfies the four conditions $AXA = A$, $XAX = X$, $(AX)^t = AX$, and

$(XA)^t = XA$, is ***always unique***, and ***always exists*** for any real or complex $m \times n$ matrix A.

The equation $Ax = b$ will have either no solution (contradictory/ inconsistent equations) or just one solution (nonsingular A) or infinity of solutions (consistent with singular A). The equation cannot have just two or just three or just k (k is any finite positive integer) solutions. For, if these have then a linear combination of these finite number of solutions is also a solution implying the existence of infinity of solutions. For example, the system of equations $3x_1 + 7x_2 = 10$, $6x_1 + 14x_2 = 19$ has no solution, i.e., we will never be able to find a numerical value of x_1 and that of an x_2 which will satisfy both the equations simultaneously. Geometrically, these two equations will represent two one-dimensional hyperplanes which are here infinite straight lines and which are non-coincident parallel lines, i.e., these two lines will never intersect implying no solution. The nature never gives rise to such a contradictory (wrong) system as a mathematical model. If 19 is replaced by 20 in the foregoing system, then we will have infinity of solutions – one solution is $x_1 = 1$, $x_2 = 1$ while another solution is $x_1 = 0$, $x_2 = 10/7$. Geometrically, the later two equations represent two coincident 1-dimensional hyperplanes implying infinity of points of intersections, i.e., infinity of solutions. If we have the equations $3x_1 + 7x_2 = 10$ and $5x_1 + 7x_2 = 12$ then there is only one solution, viz., $x_1 = 1$, $x_2 = 1$. Geometrically, these two equations will represent two non-parallel straight lines, each of infinite length, that intersect at the point (1,1). For linear systems $A\mathbf{x} = \mathbf{b}$ we will compute a solution $\mathbf{x}$ along with an error as well as the inverse A^+ also with an error.

Need for norm for error computation in matrices The matrix inverse has $n \times m$ elements and the solution vector has n elements when A in the system $A\mathbf{x} = \mathbf{b}$ is an $m \times n$ matrix. Associated with each element there is an error. Are we then going to compute $n \times m$ errors as well as n errors, respectively? The answer is certainly 'no'. From human psychology point of view, we are able to compare two values at a time and can say that one is larger than the other or both are equal. If there are two different methods/algorithms to compute A^+ and if we compute $n \times m$ errors corresponding to $n \times m$ elements of A^+ for each method then we will not be able to compare and say which method has given better result, i.e., which method has produced less errors unless we introduce some way to produce a single figure/value for errors for each method and then compare these two values to conclude logically that one is better than the other. A way to produce a single figure/value is to compute a norm of a matrix or a vector. There are different kinds of norms – Euclidean norm, spectral norm, maximum row-sum(L_∞-) norm, maximum column-sum (L_1-) norm (Krishnamurthy and Sen 2001). We will restrict ourselves to Euclidean norm which is also known as the Erhard-Schmidt norm or the Schur norm or the Frobenius norm. The

Euclidean norm (the more often used term) for an $m \times n$ matrix A is defined here by the real number

$$||A|| = (\Sigma\Sigma\, |a_{ij}|^2)^{1/2},$$

where the first summation will run over i = 1 to m and the second summation from j = 1 to n. Similarly the Euclidean norm for an n-dimensional vector $\mathbf{x} = [x_j]$ is

$$||\mathbf{x}|| = (\Sigma|x_j|^2)^{1/2},$$

where the summation will run from j =1 to j = n.

A quadratic iterative scheme for A^+ A simple iterative method with quadratic convergence to compute the minimum norm least squares inverse A^+ for a numerically specified $m \times n$ matrix A is as follows (denoting by the superscript t the transpose, by tr the trace, and by I the unit matrix of order m).

Define $X_0 = A^t/\text{tr}(AA^t)$.
Compute $X_{k+1} = X_k (2I - AX_k)$,
$k = 0, 1, 2, \ldots$, till $||X_{k+1} - X_k||/||X_{k+1}|| \le 5 \times 10^{-5}$.

The matrix X_{k+1} will be the required A^+ correct up to 4 significant digits after the execution of the foregoing method. Observe that the foregoing iterative scheme for the p-inverse A^+ will always converge with the above mentioned choice of X_0 (Sen and Prabhu 1976, Krishnamurthy and Sen 2001).

Let us compute the minimum norm least squares inverse A^+ for the given matrix A, where trace (AA^t)=17, I is the unit matrix of order 2, and

$$A = \begin{bmatrix} 3 & 1 & 1 \\ 1 & 2 & -1 \end{bmatrix}.$$

We may obtain X_0 using the Matlab command >> X=A'/trace(A*A') and X_1 using the Matlab command >>I=eye(2); X=X*(2*I – A*X):

$$X_0 = \begin{bmatrix} 0.1765 & 0.0588 \\ 0.0588 & 0.1176 \\ 0.0588 & -0.0588 \end{bmatrix}$$

$$X_1 = \begin{bmatrix} 0.2249 & 0.0554 \\ 0.0519 & 0.1799 \\ 0.0934 & -0.1107 \end{bmatrix}.$$

The realative error $||X_1 - X_0||/||X_1|| = 0.3402 > 5 \times 10^{-5}$. The successive norms $||X_{k+1} - X_k||/||X_{k+1}||$ for k = 1, 2, 3, 4, 5, and 6 are 0.3767, 0.2676, 0.1178, 0.0175, 3.1750×10^{-4}, and 1.0087×10^{-7} where the last norm satisfies the condition, viz., $1.0087 \times 10^{-7} < 5 \times 10^{-5}$. Therefore,

$$X_7 = A^+ = \begin{bmatrix} 0.2800 & -0.0200 \\ -0.0400 & 0.3600 \\ 0.2000 & -0.3000 \end{bmatrix}$$

is the required minimum norm least squares inverse correct up to 4 significant digits. Thus the relative error in each element is less than 5×10^{-5}. We have only retained four digits after the decimal point although the computation was carried out with 15 digits in the mantissa. If the vector $\mathbf{b} = [7 \quad 1]^t$ in the equation $A\mathbf{x} = \mathbf{b}$, where A is the foregoing matrix, then a solution of the consistent system is $\mathbf{x} = A^+\mathbf{b} = [1.9400 \quad 0.0800 \quad 1.1000]^t$, taking the arbitrary vector $\mathbf{z} = \mathbf{0}$ in the general solution.. Out of infinite possible solutions, this solution has the minimum norm. If we take, in the equation $A\mathbf{x} = \mathbf{b}$, $\mathbf{b} = [6 \quad 2.8]^t$ and

$$A = \begin{bmatrix} 3 & 2 & 1 \\ 1.5 & 1 & .5 \end{bmatrix}$$

we get an inconsistent system of equations. The least-squares solution (whose norm is also minimum) of this inconsistent system is $x = [1.2686 \quad 0.8457 \quad 0.4229]^t$. This solution will not satisfy the equation as the equation has no solution because of inconsistency. But the sum of the squares of the residuals, viz., $||Ax - b||^2$ is a minimum as well as the norm of the vector x, viz., $||x||$ is also a minimum. ***The minimum norm least squares solution*** x ***as well as the minimum norm least squares inverse*** A^+ ***are both unique. These are very useful in solving linear least-squares problems which arise in many physical problems including time-series analysis.***

5.5.4 Error in x of Ax=b in noniterative algorithms with nonsingular A

Consider the linear system $A\mathbf{x} = \mathbf{b}$, where A is nonsingular. It may be seen **that** ***the nonsingularity of A mathematically implies that (i) the matrix A is square, (ii) it has all the rows linearly independent as well as all the columns linearly independent, (iii) the equation Ax = b is consistent, and (iv) Ax = b has a unique solution***. The nonsingularity of A also implies that the homogenious equation $A\mathbf{x} = \mathbf{0}$ corresponding to the nonhomogeneous equation A**x** = **b** has only the trivial solution **x** = **0**. This statement is the ***fundamental theorem of linear algebra***.

Let X be an approximate inverse of the matrix A and **z** = X**b** be the approximate solution vector of the system A**x** = **b**. Choose that residual matrix Y out of the right-hand side residual matrix Y = I – AX and the left-hand side residual matrix Y = I – XA for which ||Y|| is smaller. Observe that the residual Y will be an n × n null matrix if A is n × n nonsingular and X is the exact (true) inverse of A. Let **r** = **b** – A**z** be the residual vector. If ||Y|| < 1, the ***absolute error*** (Fitzgerald 1970; Krishnamurthy and Sen 2001; Sen 2002) in the approximate solution vector **z** can be given by

$$\|\mathbf{r}\|/\|A\| \le \|A^{-1}\mathbf{b} - \mathbf{z}\| \le (\|X\| \times \|Y\| \times \|\mathbf{b}\|)/(1 - \|Y\|).$$

The leftmost term $\|\mathbf{r}\|/\|A\| = E_{min}$, say, indicates that the absolute error (always computed as a nonnegative quantity) in the computed solution vector **z** is not less than the value $\|\mathbf{r}\|/\|A\|$. The rightmost term $(\|X\| \times \|Y\| \times \|\mathbf{b}\|)/(1 - \|Y\|) = E_{max}$, say, on the other hand, denotes that the absolute error in **z** is certainly not greater than the value of this term.

Consider the linear system A**x** = **b**, where

$$A=\begin{bmatrix}1 & 2 & 3\\ 4 & 5 & 6\\ 7 & 8 & 8\end{bmatrix} \quad b=\begin{bmatrix}6\\ 15+10^{-4}\\ 23\end{bmatrix}.$$

Let the approximate (computed) inverse of the matrix A be X and the computed solution vector be **z**, where

$$X=\begin{bmatrix}-2.6667 & 2.6667 & -1.0000\\ 3.3333 & -4.3333 & 2.0000\\ -1.0000 & 2.0000 & -1.0000\end{bmatrix}, \quad \mathbf{z} = X\mathbf{b} = \begin{bmatrix}1.0003\\ 0.9996\\ 1.0002\end{bmatrix}.$$

The residual vector $\mathbf{r} = \mathbf{b} - A\mathbf{z} = 10^{-3}\,[-0.0000 \quad -0.1000 \quad 0.0000]^t$. The right-hand side residual matrix Y = I – AX, where I is the unit matrix of order 3, is

$$Y = I - AX = 10^{-3} \times \begin{bmatrix}0.0000 & -0.0000 & 0.0000\\ 0.10000 & -0.2000 & 0.1000\\ 0.0000 & 0.0000 & 0.0000\end{bmatrix}.$$

Hence $E_{min} = 6.1310 \times 10^{-6}$ while $E_{max} = 0.0509$.

The ***relative error*** (Fitzgerald 1970; Krishnamurthy and Sen 2001; Sen 2002) in **z** is, when $\|Y\| < 1$, given as $\|A^{-1}\mathbf{b} - \mathbf{z}\|/\|A^{-1}\mathbf{b}\| \le \|Y\|$. Hence the relative error in **z** is $\|Y\| = 2.4495 \times 10^{-4}$. Both the foregoing errors could be used for any ***noniterative*** algorithm. For fixed-point iterative algorithms, both the relative as well as the absolute errors are obtained just by considering the most recent solution vector and the one just preceding it.

However, if the matrix A is near-singular (i.e., the determinant of A is near-zero or, equivalently, the rows (or columns) of A are nearly linearly dependent) then the relative error will be large. Consequently, the quality of the result may not be good. However, this quality depends on the precision of the computer used. If the mathematical model $A\mathbf{x} = \mathbf{b}$ for a physical problem is

$$\overset{A}{\begin{bmatrix} 1 & 2 & 3 \\ -2 & -4 & -5+10^{-6} \\ 2 & 4+10^{-6} & 5 \end{bmatrix}} \overset{\mathbf{x}}{\begin{bmatrix} x_1 \\ x_2 \\ x_3 \end{bmatrix}} = \overset{\mathbf{b}}{\begin{bmatrix} 6 \\ -11+10^{-6} \\ 11+10^{-6} \end{bmatrix}}$$

where the determinant of A is $-1.000001000139778 \times 10^{-6}$, then the computed solution vector is

$$\mathbf{z} = \begin{bmatrix} 1.00000000372529 \\ 0.99999999627471 \\ 1.00000000000000 \end{bmatrix}$$

when computations are done with 15 digit floating-point arithmetic. The linear system is so constructed that the exact solution vector is $\mathbf{x} = [1 \quad 1 \quad 1]^t$. Thus the foregoing computation is sufficiently good. If the computations are done with only eight significant digits then the error would have been much more pronounced and possibly unacceptable.

5.5.5 Error in inverse X of nonsingular A in noniterative algorithms

Let $E = AX - XA$. If $\|Y\| < 1$, the ***absolute error*** in the ***approximate inverse*** X is given by (Fitzgerald 1970; Krishnamurthy and Sen 2001; Sen 2002)

$$\|E\|/(2 \times \|A\|) \le \|A^{-1} - X\| \le \|X\|.\|Y\|/(1 - \|Y\|).$$

The ***relative error*** in X, when $\|Y\| < 1$, is given as

$$(\|E\|/(2\|A\|))((1 - \|Y\|)/\|X\|) \le \|A^{-1} - X\|/\|A^{-1}\| \le \|Y\|(1 + \|Y\|)/(1 - \|Y\|).$$

We choose $\|Y\| = \|I - AX\|$ or $\|I - XA\|$, whichever is smaller.

Consider the linear system $A\mathbf{x} = \mathbf{b}$, where

$$A=\begin{bmatrix} 5 & 3 & 1 \\ 1 & 3 & -6 \\ 10 & 6 & 2.001 \end{bmatrix}, \quad \mathbf{b}=\begin{bmatrix} 9 \\ -2 \\ 18.001 \end{bmatrix},$$

$$X=10^3\begin{bmatrix} 3.50025000000039 & -0.00025000000000 & -1.75000000000019 \\ -5.166775000000057 & 0.00041666666667 & 2.58333333333362 \\ -2.0000000000022 & 0 & 1.00000000000011 \end{bmatrix}.$$

Since $\|Y2\| = \|I - XA\| = 4.067383956680332 \times 10^{-12} < \|Y1\| = \|I - AX\| = 7.990155261101264 \times 10^{-12}$, we choose Y = Y2. Hence the absolute error in the approximate (computed) inverse X lies in $[3.037009355793932 \times 10^{-13}, 2.980123316797209 \times 10^{-8}]$. The relative error in X lies in $[4.145024154007144 \times 10^{-17}, 4.067383956713420 \times 10^{-12}]$.

In fact, the maximum absolute error (or, simply the absolute error) and the maximum relative error (or, simply the relative error) in X are $2.980123316797209 \times 10^{-8}$ and $4.067383956713420 \times 10^{-12}$, respectively. The computed vector $\mathbf{z} = X\mathbf{b}$ is given as

$$\mathbf{z}=\begin{bmatrix} 0.99999999999636 \\ 1.00000000000728 \\ 1.00000000000000 \end{bmatrix}.$$

The vector $\mathbf{r} = \mathbf{b} - A\mathbf{z}$ is computed as

$$\mathbf{r} = 10^{-10} \times \begin{bmatrix} -0.03637978807092 \\ -0.18189894035459 \\ -0.07275957614183 \end{bmatrix}.$$

The absolute error in the computed solution vector is $6.027191666039631 \times 10^{-7}$ while the relative error is $4.067383956713420 \times 10^{-12}$. All these results depict that the computation of the approximate inverse X and the solution vector **z** is excellent in 15 digit precision. However, in lower precision, say 7 digit precision, the foregoing errors will be significantly pronounced.

Thus the higher the degree of ill-conditioning is, the higher should be the matching precision of computation so that we get a reasonable accuracy of the solution that can be acceptable or meaningfully usable in the real world environment. In most situations available to an application programmer or a user, the choice of precision is either nonexistent (e.g., MATLAB has a fixed precision of 15 digits) or very few (e.g., FORTRAN provides single, double, and sometimes quadruple precisions). Observe that higher the precision is, the more will be the computation time and one should not simply go for precision higher than what is required in a specified physical context. For a large compute-intensive problem, the amount of time for computation does matter. In situations where the time does not matter, one need not worry about using higher precision since such a usage only would take more time for computation providing harmless additional accuracy in the solution.

5.5.6 Error in x of Ax = b in noniterative algorithms with singular A

Consider the linear system $A\mathbf{x} = \mathbf{b}$, where A is singular. The singularity of the matrix A mathematically implies that (i) A is a square matrix with determinant 0 with no true inverse or (ii) A is a nonsquare rectangular matrix (also with no true inverse) which when appended with appropriate number of zero rows or zero columns produces a square matrix whose determinant is 0. The system may be consistent with infinite solutions or inconsistent (i.e., with no solution or a nonunique/unique least-squares solution or the unique minimum-norm least-squares solution). Let $\mathbf{z}$ be an approximate solution of the system $A\mathbf{x} = \mathbf{b}$ with $\mathbf{b} \neq \mathbf{0}$ (null column vector). Let $\Delta\mathbf{b} = AA^+\mathbf{b} - \mathbf{b}$. The ***relative error*** (Sen 2002; Sen et al. 2000) for the minimum norm least squares solution $\mathbf{z} = A^+\mathbf{b}$ of the consistent/inconsistent system $A\mathbf{x} = \mathbf{b}$ is defined as

$$||\mathbf{x} - \mathbf{z}||/||\mathbf{x}|| \leq (||(I - AA^+)\Delta\mathbf{b}|| + ||A^+\Delta\mathbf{b}||)/||A^+(\mathbf{b} + \Delta\mathbf{b})||.$$

We need to compute the right-hand side of the inequation to obtain the relative error for the solution vector.

The ***inconsistency index*** (Sen 2002; Sen et al. 2000) of the system $A\mathbf{x} = \mathbf{b}$ (consistent or inconsistent) is

$$Inci = ||\Delta\mathbf{b}||/||A, \mathbf{b}||$$

Consider the near-consistent (strictly inconsistent) system $A\mathbf{x} = \mathbf{b}$, where

$$A=\begin{bmatrix}5 & 3 & 2\\ 10 & 6 & 4\end{bmatrix},\ \mathbf{b}=\begin{bmatrix}10\\ 20.01\end{bmatrix},\ A^{+}=\begin{bmatrix}.0286 & .0571\\ .0171 & .0343\\ .0057 & .0014\end{bmatrix},\ \Delta\mathbf{b}=\begin{bmatrix}.0040\\ -.0020\end{bmatrix},$$

$$A^{+}=\begin{bmatrix}0.02631578947368 & 0.05263157894737\\ 0.01578947368421 & 0.03157894736842\\ 0.01052631578947 & 0.02105263157895\end{bmatrix}$$

$$\Delta\mathbf{b}=\begin{bmatrix}0.00400000000000\\ -0.00200000000000\end{bmatrix}.$$

The relative error is 0.00275570746743 while the inconsistency index *Inci* = 1.702019677445941 $\times$ 10^{-4}. The relative error does take into account the slight inconsistency that exists in the linear system. The foregoing inconsistency index indicates that the system is not very inconsistent. If the *Inci* is 0 then the system is completely consistent. If it is small then it may be termed near-consistent. If it is large then it is highly inconsistent.

In the event of a highly inconsistent system, it is necessary for one to go back to the real-world (physical) problem and then check thoroughly the derived mathematical model along with the assumptions (if any) made as well as the possible error (order of error or error bound) introduced due to the inherent error existing in a measuring device. It is necessary to eliminate all possible human errors. It is also necessary to appropriately modify the assumptions (if any) so that these assumptions are much more reasonable and do not contribute to the inconsistency beyond an acceptable limit. In fact, the resulting mathematical model must not be highly inconsistent (implying serious contracdiction inside the system). Observe that the real-world problems or the problems in nature are always consistent (rather 100% consistent); in other words, inconsistency is completely unknown in nature. In fact, inconsistency is always a creation by human beings.

5.5.7 Iterative algorithms for linear system

Let A**x** = **b** be the given linear system, where the matrix A is n×n. Let A = L + U + D, where L = the strictly lower triangular matrix (i.e., the lower triangular matrix whose diagonal elements are all zero), U = the strictly upper triangular matrix (i.e., the upper triangular matrix whose diagonal elements are all zero), and D = the diagonal matrix. If

$$A=\begin{bmatrix}5 & 3 & 2\\ 2 & -7 & 11\\ 6 & 8 & 4\end{bmatrix},\ L=\begin{bmatrix}0 & 0 & 0\\ 2 & 0 & 0\\ 6 & 8 & 0\end{bmatrix},\ U=\begin{bmatrix}0 & 3 & 2\\ 0 & 0 & 11\\ 0 & 0 & 0\end{bmatrix},\ D=\begin{bmatrix}5 & 0 & 0\\ 0 & -7 & 0\\ 0 & 0 & 4\end{bmatrix}.$$

For the sake of convergence, we may interchange the rows and columns of the augmented matrix (A, **b**) so that the matrix becomes diagonally dominant. We do such row/column permutations for the following iterative algorithms. Consider the system $A\mathbf{x} = \mathbf{b}$, where

$$A=\begin{bmatrix}5 & 3 & 2\\ 2 & -7 & 11\\ 6 & 8 & 4\end{bmatrix},\ \mathbf{b}=\begin{bmatrix}17\\ 21\\ 34\end{bmatrix}.$$

To make the diagonal elements largest, we interchange the first and the second rows and then the first and the third columns of the augmented matrix (A, **b**). The column interchange necessitates the interchange of the elements x_1 and x_3 of the solution vector **x**. Observe that the row interchanges do not induce any interchange in the elements of the solution vector **x**. Finally we interchange the second and third rows of the augmented matrix. The resulting system is now

$$\overset{A}{\begin{bmatrix}11 & -7 & 2\\ 4 & 8 & 6\\ 2 & 3 & 5\end{bmatrix}}\ \overset{x}{\begin{bmatrix}x_3\\ x_2\\ x_1\end{bmatrix}} = \overset{b}{\begin{bmatrix}21\\ 34\\ 17\end{bmatrix}}.$$

In the ***Jacobi method*** (Krishnamurthy and Sen 2001; Sen 2002) we write the iteration as $\mathbf{x}^{(k+1)} = D^{-1}(\mathbf{b} - (L + U)\mathbf{x}^{(k)})$. If we select $\mathbf{x}^{(0)} = [x_3\ \ x_2\ \ x_1]^t = [2\ \ 2\ \ 2]^t$ then the successive iterates are

$$\mathbf{x}^{(1)} = [x_3\ \ x_2\ \ x_1]^{(1)t} = [2.81818181818182\ \ 1.75000000000000\ \ 1.40000000000000]^t,$$

$$\mathbf{x}^{(2)} = [x_3\ \ x_2\ \ x_1]^{(2)t} = [2.76818181818182\ \ 1.79090909090909\ \ 1.22272727272727]^t,$$

$$\vdots$$

$$.\mathbf{x}^{(20)} = [x_3\ \ x_2\ \ x_1]^{(20)t} = [2.99991675061462\ \ 1.99994488807044\ \ 1.00010284939062]^t.$$

$\mathbf{x}^{(21)} = [x_3 \;\; x_2 \;\; x_1]^{(21)t}$
$= [2.99994622888289 \;\; 1.99996448764972 \;\; 1.00006636691189]^t.$

The relative error in the solution vector $\mathbf{x}^{(20)} = ||\mathbf{x}^{(21)} - \mathbf{x}^{(20)}||/||\mathbf{x}^{(21)}|| = 1.358608619057123 \times 10^{-5}$. Hence the solution vector $\mathbf{x}^{(20)} = [x_3 \;\; x_2 \;\; x_1]^{(20)t}$ has each of the elements correct at least up to 4 significant digits. For most real-world problems this accuracy is good enough. Observe that the solution vector $\mathbf{x}^{(21)} = [x_3 \;\; x_2 \;\; x_1]^{(21)t}$ is still more accurate. The actual computation was done using a precision of 15 digits.

In the ***Gauss-Seidel method*** (Krishnamurthy and Sen 2001; Sen 2002), we write the iteration as $\mathbf{x}^{(k+1)} = (L + D)^{-1} (\mathbf{b} - U\mathbf{x}^{(k)})$. For the foregoing example with the same L, U, D, and initial approximation $\mathbf{x}^{(0)}$ as in the Jacobi method, we obtain

$\mathbf{x}^{(1)} = [x_3 \;\; x_2 \;\; x_1]^{(1)t}$
$= [2.81818181818182 \;\; 1.34090909090909 \;\; 1.46818181818182]^t,$

$\mathbf{x}^{(2)} = [x_3 \;\; x_2 \;\; x_1]^{(2)t}$
$= [2.49545454545455 \;\; 1.90113636363636 \;\; 1.26113636363636]^t,$

$\vdots$

$\mathbf{x}^{(15)} = [x_3 \;\; x_2 \;\; x_1]^{(15)t}$
$= [2.99993527825892 \;\; 1.99996084722832 \;\; 1.00004938035944]^t,$

$\mathbf{x}^{(16)} = [x_3 \;\; x_2 \;\; x_1]^{(16)t}$
$= [2.99996610635267 \;\; 1.99997991155409 \;\; 1.00002561052648]^t.$

The relative error in the solution vector $\mathbf{x}^{(15)}$ is $||\mathbf{x}^{(16)} - \mathbf{x}^{(15)}||/||\mathbf{x}^{(16)}|| = 1.158464513827630 \times 10^{-5} < 5 \times 10^{-5}$ Hence the required solution is $\mathbf{x} = [x_3 \;\; x_2 \;\; x_1]^t = [2.9999 \;\; 1.9999 \;\; 1.0001]^t$, which is correct at least up to 4 significant digits. Observe that the solution vector $\mathbf{x}^{(16)}$ is more accurate than $\mathbf{x}^{(15)}$ and the Gauss-Seidel method converges faster than the Jacobi method. In the foregoing example it has taken only 15 iterations instead of 20 (taken by the Jacobi iteration). Over- or under-relaxation of the Gauss-Seidel iteration does not help much for this problem. If we write the ***relaxation scheme*** as

$$\mathbf{x}^{(k+1)} = (D + \omega L)^{-1}[D\mathbf{x}^{(k)} - \omega\{(U + D)\mathbf{x}^{(k)} - \mathbf{b}\}]$$

and take $\omega = 1.9$ then the scheme will diverge. If we choose $\omega = 0.7$, then the scheme will converge slower than the Gauss-Seidel method and even the

Jacobi method. The relaxation scheme will become the Gauss-Seidel scheme when the relaxation factor $\omega = 1$.

Linear partial differential equations with initial/boundary conditions are often approximated by linear finite difference equations which are solved either noniteratively using the Gauss Reduction method with partial pivoting (Krishnamurthy and Sen 2001; Sen 2002) or iteratively by the Gauss-Seidel/relaxation method. The errors are computed as shown earlier to logically establish the quality of the result.

The Jacobi, the Gauss-Seidel, as well as the relaxation methods (iterative) can also be developed for a ***nonlinear system of equations*** (Krishnamurthy and Sen 2001) and these are useful for solving nonlinear partial differential equations when approximated by finite difference equations (which are nonlinear). The errors are also computed straightway since (when the methods have sufficiently converged) the solution vector $\mathbf{x}$ of higher-order accuracy as well as that of lower-order accuracy are known from the successive iterations.

5.5.8 Eigenvalues and eigenvectors of a matrix

The eigenvalues and eigenvectors are defined for an $n \times n$ (singular or nonsingular) matrix A and not for an $m \times n$ rectangular matrix, where $m \neq n$.. If A is nonsquare then we may append appropriate number of zero rows or zero columns to make it square before we talk about its eigenvalues and eigenvectors. If we find a scalar λ and an $n \times 1$ vector $\mathbf{x}$ so that $A\mathbf{x} = \lambda\mathbf{x}$ then λ is one of n possible eigenvalues and $\mathbf{x}$ is the corresponding eigenvector of A. The eigenvalue computation is identical with the problem of computing the roots of a polynomial equation while the eigenvector computation is identical with that of computing a solution vector $\mathbf{x}$ of the linear homogeneous equation $(A - \lambda I)\mathbf{x} = \mathbf{0}$ (null column vector), where I is the $n \times n$ unit matrix and the eigenvalue λ is known/computed. The eigenvalues will be computed iteratively while the eigenvectors are obtained in the process of this iteration or by actually solving the homogeneous equation noniteratively or iteratively.

5.5.9 Power method for an eigenvalue and an eigenvector

There are many methods (Krishnamurthy and Sen 2001; Sen 2002) to compute the n eigenvalues (counting multiplicities) and the corresponding n eigenvectors of a given $n \times n$ real or complex (numerical) matrix A. Consider an iterative method, called the power method, to compute the largest magnitude and then the smallest magnitude eigenvalues and the corresponding eigenvectors of A whose eigenvalues are $\lambda_1, \lambda_2, \ldots, \lambda_n$, where $|\lambda_1| > |\lambda_2| \geq |\lambda_3| \geq \ldots \geq |\lambda_n|$.

- Choose the initial vector $\mathbf{x}^{(0)} = [1\ \ 1\ \ldots\ 1]^t$.
- Compute $\mathbf{y}^{(p+1)} = A\mathbf{x}^{(p)}$, $\beta_{p+1} = \max_k |(\mathbf{y}^{(p+1)})_k|$, $\mathbf{x}^{(p+1)} = \mathbf{y}^{(p+1)}/\beta_{p+1}$,

p=0, 1, . . ., till $||\mathbf{x}^{(p+1)} - \mathbf{x}^{(p)}||/||\mathbf{x}^{(p+1)}|| \le 5 \times 10^{-5}$ (for 4 significant digit accuracy).

The value $\beta_{p+1} = \lambda_1$ and the vector $\mathbf{x}^{(p+1)}$ will give the largest magnitude eigenvalue and the corresponding eigenvector (in standardized form), respectively. A nonzero multiple of an eigenvector is also an eigenvector. The standardized form of an eigenvector is one in which the largest element is 1. Having thus computed the largest magnitude eigenvalue λ_1, the smallest magnitude eigenvalue (distinct) is computed using the power method for the matrix $(A - \lambda_1 I)$ instead of the matrix A. The matrix has the eigenvalues $\lambda'_k = (\lambda_k - \lambda_1)$, k = 1, 2, . ., n. λ'_n is evidently the largest magnitude eigenvalue of $(A - \lambda_1 I)$, which is computed using the power method. Consider the matrix A (Sen 2002) with the initial approximation $\mathbf{x}^{(0)} = [1\ \ 1]^t$, where

$$A = \begin{bmatrix} 1 & 2 \\ 3 & 4 \end{bmatrix},$$

p=0, $\mathbf{y}^{(1)} = A\mathbf{x}^{(0)} = [3\ \ 7]^t$,
$\beta_1 = \max |\mathbf{y}^{(1)}| = 7$,
$\mathbf{x}^{(1)} = \mathbf{y}^{(1)}/\beta_1 = [0.42857142857143\ \ 1.00000000000000]^t$;

p=1, $\mathbf{y}^{(2)} = A\mathbf{x}^{(1)} = [2.42857142857143\ \ 5.28571428571429]^t$,
$\beta_2 = \max|\mathbf{y}^{(2)}| = 5.28571428571429$,
$\mathbf{x}^{(2)} = \mathbf{y}^{(2)}/\beta_2 = [0.45945945945946\ \ 1.00000000000000]^t$;

p = 2, $\mathbf{y}^{(3)} = A\mathbf{x}^{(2)} = [2.45945945945946\ \ 5.37837837837837]^t$,
$\beta_3 = \max|\mathbf{y}^{(3)}| = 5.37837837837837$,
$\mathbf{x}^{(3)} = \mathbf{y}^{(3)}/\beta_3 = [0.45728643216080\ \ 1.00000000000000]^t$;

p = 3, $\mathbf{y}^{(4)} = A\mathbf{x}^{(3)} = [2.45728643216081\ \ 5.37185929648242]^t$,
$\beta_4 = \max|\mathbf{y}^{(4)}| = 5.37185929648242$,
$\mathbf{x}^{(4)} = \mathbf{y}^{(4)}/\beta_4 = [0.45743685687558\ \ 1.00000000000000]^t$;

p = 4, $\mathbf{y}^{(5)} = A\mathbf{x}^{(4)} = [2.45743685687558\ \ 5.37231057062675]^t$;
$\beta_5 = \max|\mathbf{y}^{(5)}| = 5.37231057062675$,
$\mathbf{x}^{(5)} = \mathbf{y}^{(5)}/\beta_5 = [0.45742643217830\ \ 1.00000000000000]^t$;

p = 5, $\mathbf{y}^{(6)} = A\mathbf{x}^{(5)} = [2.45742643217830\ \ 5.37227929653491]^t$;
$\beta_6 = \max|\mathbf{y}^{(6)}| = 5.37227929653491$,
$\mathbf{x}^{(6)} = \mathbf{y}^{(6)}/\beta_6 = [0.45742715457168\ \ 1.00000000000000]^t$;

p = 6, $\mathbf{y}^{(7)} = A\mathbf{x}^{(6)} = [2.45742715457168 \quad 5.37228146371504]^t$;
$\beta_7 = \max|\mathbf{y}^{(7)}| = 5.37228146371504$,
$\mathbf{x}^{(7)} = \mathbf{y}^{(7)}/\beta_7 = [0.45742710451219 \quad 1.00000000000000]^t$.

The relative error $e_r = ||\mathbf{x}^{(7)} - \mathbf{x}^{(6)}||/||\mathbf{x}^{(7)}|| = 4.552293347263694 \times 10^{-8} < 5 \times 10^{-5}$.

Hence the largest magnitude eigenvalue is $\lambda_1 = \beta_6 = 5.37227929653491$ which is correct up to 4 significant digits and the corresponding eigenvector (in standardized form) is $\mathbf{x} = \mathbf{x}^{(6)} = [0.45742715457168 \; 1.00000000000000]^t$ which is also correct up to 4 significant digits. Observe that the largest magnitude eigenvalue β_7 as well as the corresponding eigenvector $\mathbf{x} = \mathbf{x}^{(7)}$ are still more accurate than the preceding eigenvalue and the eigenvector.

5.5.10 Linear programming – Karmarkar algorithm

To discuss about error in an algorithm or to compute the error in it, it is necessary to specify the algorithm. To obtain an error estimate in the Karmarkar algorithm for linear program (LP), which is iterative we first write down the formal steps of the algorithm preceding by the statement of the Karmarkar form of linear programs (KLP) which is not the same as that of an LP in equality/inequality constraints.

A linear program (LP) is defined as *Min (Minimize)* $z = \mathbf{c}^t\mathbf{x}$ *subject to* $A\mathbf{x} \leq \mathbf{b}$, $\mathbf{x} \geq \mathbf{0}$ (null column vector), where $A=[a_{ij}]$ is an $m \times n$ numerically specified matrix, $\mathbf{b}=[b_i]$ is an $m \times 1$ numerically given column vector and $\mathbf{c} = [c_j]$ is an $n \times 1$ numerically specified column vector.

Let $\mathbf{e}$ be the vector $[1 \; 1 \; . \, . \, . \; 1]^t$ of appropriate order. The Karmarkar form of linear program (KLP), on the other hand, is defined as *Min* $z=\mathbf{c}^t\mathbf{x}$ *subject to* $A\mathbf{x}=\mathbf{0}$, $\mathbf{e}^t\mathbf{x}=1$, $\mathbf{x}\geq\mathbf{0}$, $\mathbf{x}=\mathbf{e}/n$ *is feasible, minimal* z *value* = 0.

The Karmarkar algorithm (KA) uses a projective transformation f to create a set of transformed variables $\mathbf{y}$ (Karmarkar 1984; Sen 2002). f always transforms the current point $\mathbf{x}$ into the centre of the feasible region in the space defined by the transformed variables. If f takes the point $\mathbf{x}$ into the point $\mathbf{y}$, then we write $f(\mathbf{x}) = \mathbf{y}$. The KA begins in the transformed space in a direction that tends to improve the objective function value z without violating feasibility. This yields a point $\mathbf{y}^1$, close to the boundary of the feasible region, in the transformed space. The new point is $\mathbf{x}^1$ that satisfies $f(\mathbf{x}^1) = \mathbf{y}^1$. The procedure is iterated replacing $\mathbf{x}^0$ by $\mathbf{x}^1$ until z for $\mathbf{x}^r$ is sufficiently small (close to 0).

Conversion of LP to KLP One needs to rewrite/convert the foregoing conventional linear program (LP) or the LP in standard form (constraints are equations) to Karmarkar form of linear program (KLP) before one could use the Karmarkar algorithm (Sen 2002). Let $\mathbf{s}$ and $\mathbf{v}$ be the vectors $[s_j] = [s_1$

$s_2 \ldots s_m]^t$ of slack variables and $[v_i] = [v_1 \;\; v_2 \ldots v_n]^t$ of surplus variables, respectively. Consider the LP

Maximize (Max) $z = \mathbf{c}^t\mathbf{x}$ *subject to* $A\mathbf{x} \le \mathbf{b}$, $\mathbf{x} \ge \mathbf{0}$ (null column vector).

The dual of this LP is

Minimize (Min) $w = \mathbf{b}^t\mathbf{y}$ *subject to* $A^t\mathbf{y} \ge \mathbf{c}$, $\mathbf{y} \ge \mathbf{0}$.

If the solution $\mathbf{x}$ is feasible in the LP, the solution $\mathbf{y}$ is feasible in the dual of LP, and $z = w$, then $\mathbf{x}$ is maximal for LP (Duality theorem). Hence any feasible solution of the constraints

$$\mathbf{c}^t\mathbf{x} - \mathbf{b}^t\mathbf{y} = \mathbf{0},\;\; A\mathbf{x} \le \mathbf{b},\; A^t\mathbf{y} \ge \mathbf{c},\; \mathbf{x}, \mathbf{y} \ge \mathbf{0}$$

will produce the maximal $\mathbf{x}$. Inserting the slack and surplus variables, we get the equations

$$\mathbf{c}^t\mathbf{x} - \mathbf{b}^t\mathbf{y} = \mathbf{0},\;\; A\mathbf{x} + I_m\mathbf{s} = \mathbf{b},\; A^t\mathbf{y} - I_n\mathbf{v} = \mathbf{c},\;\; (\mathbf{x}, \mathbf{y}, \mathbf{s}, \mathbf{v} \ge \mathbf{0}),$$

where I_m and I_n are the identity matrices of order m and n, respectively.

Append the equality constraint

$$\mathbf{e}^t\mathbf{x} + \mathbf{e}^t\mathbf{y} + \mathbf{e}^t\mathbf{s} + \mathbf{e}^t\mathbf{v} + d_1 = k$$

with the foregoing equations, where the value of k should be chosen such that the sum of the values of all the variables $\le k$ and the variable $d_1 \ge 0$ is dummy (slack). Thus, we have

$$\begin{gathered}\mathbf{c}^t\mathbf{x} - \mathbf{b}^t\mathbf{y} = \mathbf{0},\; A\mathbf{x} + I_m\mathbf{s} = \mathbf{b},\;\; A^t\mathbf{y} - I_n\mathbf{v} = \mathbf{c},\\ \mathbf{e}^t\mathbf{x} + \mathbf{e}^t\mathbf{y} + \mathbf{e}^t\mathbf{s} + \mathbf{e}^t\mathbf{v} + d_1 = k,\\ \mathbf{x}, \mathbf{y}, \mathbf{s}, \mathbf{v}, d_1 \ge \mathbf{0},\end{gathered}$$

Introducing the slack variable d_2 (=1) to make the nonzero right-hand sides 0, we get.

$$\begin{gathered}\mathbf{c}^t\mathbf{x} - \mathbf{b}^t\mathbf{y} = \mathbf{0},\\ A\mathbf{x} + I_m\mathbf{s} - I_m\mathbf{b}d_2 = \mathbf{0},\\ A^t\mathbf{y} - I_n\mathbf{v} - I_n\mathbf{c}d_2 = \mathbf{0},\\ \mathbf{e}^t\mathbf{x} + \mathbf{e}^t\mathbf{y} + \mathbf{e}^t\mathbf{s} + \mathbf{e}^t\mathbf{v} + d_1 - kd_2 = 0,\\ \mathbf{e}^t\mathbf{x} + \mathbf{e}^t\mathbf{y} + \mathbf{e}^t\mathbf{s} + \mathbf{e}^t\mathbf{v} + d_1 + d_2 = k + 1,\\ \mathbf{x}, \mathbf{y}, \mathbf{s}, \mathbf{v}, d_1, d_2 \ge \mathbf{0}.\end{gathered}$$

Changing the variables

$$[\mathbf{x}\;\; \mathbf{y}\;\; \mathbf{s}\;\; \mathbf{v}\;\; d_1\;\; d_2] = (k+1)[\,\mathbf{x}'\;\; \mathbf{y}'\;\; \mathbf{s}'\;\; \mathbf{v}'\;\; d_1'\;\; d_2'],$$

we write

$$\mathbf{c}^t\mathbf{x}' - \mathbf{b}^t\mathbf{y}' = \mathbf{0},$$
$$A\mathbf{x}' + I_m\mathbf{s}' - I_m\mathbf{b}d_2' = \mathbf{0},$$
$$A^t\mathbf{y}' - I_n\mathbf{v}' - I_n\mathbf{c}d_2' = \mathbf{0},$$
$$\mathbf{e}^t\mathbf{x}' + \mathbf{e}^t\mathbf{y}' + \mathbf{e}^t\mathbf{s}' + \mathbf{e}^t\mathbf{v}' + d_1' - kd_2' = 0,$$
$$\mathbf{e}^t\mathbf{x}' + \mathbf{e}^t\mathbf{y}' + \mathbf{e}^t\mathbf{s}' + \mathbf{e}^t\mathbf{v}' + d_1' + d_2' = 1,$$
$$\mathbf{x}', \mathbf{y}', \mathbf{s}', \mathbf{v}', d_1', d_2' \geq \mathbf{0}.$$

To enforce that a solution that sets all variables equal is feasible, insert the third variable d_3' to the last but one constraint and then add a multiple of d_3' to each of its preceding constraints.This insertion and addition will enforce that a solution that sets all variables equal is feasible, Choosing the multiple so that the sum of the coefficients of all variables in each constraint (except the last two) equals zero, we obtain the KLP as follows.

KLP: *Min* d_3' *subject to*

$$\mathbf{c}^t\mathbf{x}' - \mathbf{b}^t\mathbf{y}' - (\mathbf{e}^t\mathbf{c} - \mathbf{e}^t\mathbf{b})d_3' = \mathbf{0},$$
$$A\mathbf{x}' + I_m\mathbf{s}' - I_m\mathbf{b}d_2' - [A\mathbf{e} + I_m(1 - d_2')\mathbf{e}]d_3' = \mathbf{0},$$
$$A^t\mathbf{y}' - I_n\mathbf{v}' - I_n\mathbf{c}d_2' - [A^t\mathbf{e} - I_n(1 - d_2')\mathbf{e}]d_3' = \mathbf{0},$$
$$\mathbf{e}^t\mathbf{x}' + \mathbf{e}^t\mathbf{y}' + \mathbf{e}^t\mathbf{s}' + \mathbf{e}^t\mathbf{v}' + d_1' - kd_2' - (2n + 2m + 1 - k)d_3' = 0,$$
$$\mathbf{e}^t\mathbf{x}' + \mathbf{e}^t\mathbf{y}' + \mathbf{e}^t\mathbf{s}' + \mathbf{e}^t\mathbf{v}' + d_1' + d_2' + d_3' = 1,$$
$$\mathbf{x}', \mathbf{y}', \mathbf{s}', \mathbf{v}', d_1', d_2', d_3' \geq \mathbf{0}.$$

Note that the number of variables has increased. We have now totally $2m + 2n + 3$ variables. Since d_3' should be 0 in a feasible solution, we minimize d_3' in KLP. For a feasible solution, d_3' in KLP will be 0. The value of $\mathbf{x}$ in the minimal solution of KLP will produce an optimal solution of the original LP. The KLP *for the KA* can be restated, setting $m' = m + n + 3$ and $n' = 2m + 2n + 3$, as

Min $z = \mathbf{c}^t\mathbf{x}$ *subject to* $A\mathbf{x} = \mathbf{b}$,

where A is $m' \times n'$ matrix, $\mathbf{c}^t = [\mathbf{0} \;\; 1]^t$ and $\mathbf{b}^t = [\mathbf{0} \;\; 1]^t$ in which the first $\mathbf{0}$ is the $(n'-1)$ null row vector while the second $\mathbf{0}$ is the $(m'-1)$ null row vector. A, $\mathbf{x}$, and z are such that

$$\mathbf{e}^t\mathbf{x} = 1, \mathbf{x} \geq \mathbf{0},$$

and

$$\mathbf{x} = \mathbf{e}/n'$$

is feasible, minimal z-value = 0 (if the KLP is feasible).

5.5.11 Karmarkar Algorithm (KA)

If a (feasible) solution having the z-value, viz., $d_3' < \varepsilon$ (ε is a small positive value compared to the average element of A, **b**, **c**) is acceptable, then we may compute

$$\varepsilon = 5 \times 10^{-5} \times (\Sigma\Sigma|a_{ij}| + \Sigma|b_i| + \Sigma|c_j|)/(m' \times n' + m' + n')$$

for 4 significant digit accuracy, where the double summation is over $i=1(1)m'$ and $j=1(1)n'$.

The KA may now be described as follows.

S1 Input k, m′, n′, n′-vector **e**, A, **b**, and **c**. Set feasible point $\mathbf{x}^0 = \mathbf{e}/n'$ and the iterate r=0.

S2 If $(k+1)\,\mathbf{c}^t\mathbf{x}^r < \varepsilon$ then stop otherwise go to Step **S3.**

S3 Compute the new point (an n′-vector) $\mathbf{y}^{r+1}$ in the transformed n′-dimensional unit simplex S (S is the set of points **y** satisfying $\mathbf{e}^t\mathbf{y} = 1$, $\mathbf{x} \geq \mathbf{0}$), where I is the $n' \times n'$ unit matrix:

$$\mathbf{y}^{r+1} = \mathbf{x}^0 - \alpha\mathbf{c}_p/[\sqrt{(n'(n'-1))}\|\mathbf{c}_p\|],$$

where

$$\mathbf{c}_p = (I - P^t(PP^t)^+P)[\mathrm{diag}(\mathbf{x}^r)]\mathbf{c},\ \ P = \begin{bmatrix} A[\mathrm{diag}(\mathbf{x}^r)] \\ \mathbf{e}^t \end{bmatrix},\ \ 0<\alpha<1.$$

We may choose $\alpha = 0.25$. For some values of α in (0, 1) the convergence of the Karmarkar iterative algorithm is faster and for some other values it could be slower. There is no a priori simple procedure to obtain the optimal α. The matrix P is the $(m' + 1)\times n'$ matrix whose last row $\mathbf{e}^t$ is the vector of 1s. The matrix $(PP^t)^+$ is the p-inverse (Lakshmikantham et al 1996) or, equivalently, the minimum norm least-squares inverse of the matrix PP^t.

S4 Compute now the new point $\mathbf{x}^{r+1}$ in the original space using the Karmarkar centering transformation to find the point corresponding to the point $\mathbf{y}^{r+1}$:

$$\mathbf{x}^{r+1} = \mathbf{q}/(\mathbf{e}^t\mathbf{q}),\ \text{where}\ \mathbf{q} = [\mathrm{diag}(\mathbf{x}^r)]\mathbf{y}^{r+1}.$$

Set r = r + 1 and return to the step **S2**.

Note An equivalent form of $\mathbf{x}^{r+1}$ in Step **S4** is $x_j^{r+1} = x_j^r y_j^{r+1} / \Sigma x_j^r y_j^{r+1}$ j = 1(1)n′ where the summation is over t = 1(1)n′.

Consider, as a numerical example, the LP *Max* $\mathbf{c}^t\mathbf{x}$ *subject to* $A\mathbf{x} \le \mathbf{b}$, $\mathbf{x} \ge \mathbf{0}$, where

$$A = \begin{bmatrix} 1 & 2 & 1 \\ -4 & -2 & 7 \end{bmatrix}, \quad \mathbf{b} = \begin{bmatrix} 1 \\ 2 \end{bmatrix}, \quad \mathbf{c} = \begin{bmatrix} -2 \\ -7 \\ 2 \end{bmatrix}.$$

We have m = 2 and n = 3. The KLP is, choosing k = 20 and setting $x_j = 21x_j'$ j = 1(1)n, $y_i = 21y_i'$ i = 1(1)m, $s_i = 21s_i'$ i = 1(1)m, $v_j = 21v_j'$ j = 1(1)n, $d_1 = 21d_1'$, $d_2 = 21d_2'$, $\mathbf{d}' = [d_1' \;\; d_2' \;\; d_3']^t$, *Min* d_3', where all variables ≥ 0, *subject to*

$$\begin{bmatrix} -2 & -7 & 2 & -1 & -2 & 0 & 0 & 0 & 0 & 0 & 0 & 0 & 10 \\ 1 & 2 & 1 & 0 & 0 & 1 & 0 & 0 & 0 & 0 & 0 & -1 & -4 \\ -4 & -2 & 7 & 0 & 0 & 0 & 1 & 0 & 0 & 0 & 0 & -2 & 0 \\ 0 & 0 & 0 & 1 & -4 & 0 & 0 & -1 & 0 & 0 & 0 & 2 & 2 \\ 0 & 0 & 0 & 2 & -2 & 0 & 0 & 0 & -1 & 0 & 0 & 7 & -6 \\ 0 & 0 & 0 & 1 & 7 & 0 & 0 & 0 & 0 & -1 & 0 & -2 & -5 \\ 1 & 1 & 1 & 1 & 1 & 1 & 1 & 1 & 1 & 1 & 1 & -20 & 9 \\ 1 & 1 & 1 & 1 & 1 & 1 & 1 & 1 & 1 & 1 & 1 & 1 & 1 \end{bmatrix} \begin{bmatrix} \mathbf{x}' \\ \mathbf{y}' \\ \mathbf{s}' \\ \mathbf{v}' \\ \mathbf{d}' \\ \\ \\ \\ \end{bmatrix} = \begin{bmatrix} 0 \\ 0 \\ 0 \\ 0 \\ 0 \\ 0 \\ 0 \\ 1 \end{bmatrix}.$$

The foregoing KLP is the required linear program for the KA. Without any loss of generality, the general form of the KLP can be written as

Min $z = \mathbf{c}^t\mathbf{x}$ *subject to* $A\mathbf{x} = \mathbf{b}$,

where A is (m + n + 3) × (2m + 2n + 3) matrix, $\mathbf{c}^t = [\mathbf{0} \;\; 1]^t$ and $\mathbf{b}^t = [\mathbf{0} \;\; 1]^t$ in which the first **0** is the (2m + 2n + 2) null column vector while the second **0** is the (m + n + 2) null column vector. A, **x,** and z are such that $\mathbf{e}^t\mathbf{x} = 1$, $\mathbf{x} \ge \mathbf{0}$, $\mathbf{x} = \mathbf{e}/(2m + 2n + 3)$ is feasible, minimal z-value = 0 (if the KLP is feasible).

We use this general form for the KA. Set m′ = m+n+3 and n′=2m+2n+3. Hence m′ = 8, n′ = 13, A = the foregoing 8 × 13 matrix. The cost vector **c** =[0 0 0 0 0 0 0 0 0 0 0 0 1]^t. $\varepsilon = 5 \times 10^{-5} \times 149/125 = 5.96 \times 10^{-5}$ for 4 significant digit accuracy. **b** = [0 0 0 0 0 0 0 1]^t. The number of iterations for 4 significant digit accuracy is 1760. The solution is

$\mathbf{x}^{1760}$ = [0.00000529552922 0.00000070621843 0.01360668416753
0.00000584036793 0.01360902377647 0.03401706989387
0.00001588805050 0.04081503603589 0.30611828315260
0.00001589988582 0.54416762777704 0.04762003992182
0.00000260522289]^t.

Hence $x_1 = 21x'_1 = 21 \times 0.00000529552922 = 0$ (up to 4 significant digits). $x_2 = 21x'_2 = 21 \times 0.00000070621843 = 0$ (up to 4 significant digits) and $x_3 = 21x'_3 = 21 \times 0.01360668416753 = 0.28574036751813$, the objective function value $z = 2x_3 = 0.57148073503626$. Since the result is correct up to 4 significant digits, we have $\log_{10}(1/\text{relative error}) = 4$. Therefore, the relative error in the foregoing result is 0.01%. For a small LP problem the KA is too expensive (much more than the much used simplex algorithm).

5.5.12 Numerical integration – Simpson 1/3 rule

The Simpson 1/3 rule over the closed interval $[x_0, x_n]$ where $n = 2m$ = the number of subintervals must be even and the subintervals are equal, i.e., $h = (x_n - x_0)/n$, is

$$\int f(x)dx=(h/3)[f(x_0) + 4f(x_1) + 2f(x_2) + 4f(x_3) + 2f(x_4) + 4f(x_5) + \ldots + 2f(x_{2m-2}) + 4f(x_{2m-1}) + f(x_{2m})] + [(x_{2m} - x_0)/90]h^4\,|f^{(4)}(\xi)|,$$

the interval of integration being $[x_0, x_n]$, i.e., $[x_0, x_{2m}]$ and the global error being the last term, where ξ is such that $x_0 \le \xi \le x_n$ and the fourth derivative $|f^{(4)}(\xi)|$ is the largest. If the function f(x) and its derivatives are known analytically then it would be, in general, possible to compute the largest $|f^{(4)}(x)|$ in the interval $x_0 \le x \le x_n$. If f(x) is known in the form of a table $(x_i, f(x_i))$ $i = 0(1)n$, we should find numerically the largest magnitude fourth derivative of f(x) in $[x_0, x_n]$.

Consider the following *physical problem*. Let the present be $t = 0$. At time t years from now Ronald earns income at a rate of $e^{1.4t}$. Use the Simpson's 1/3 rule (with h = 1) to compute how much money ($) he earns during the next 6 years. The amount of money that Ronald earns during the next 6 years is given by the mathematical model

$$I=\int_0^6 e^{1.4t}dt$$

$$\begin{aligned} I &= [e^{1.4}\times0 + 4e^{1.4}\times1 + 2e^{1.4}\times2 + 4e^{1.4}\times3 + 2e^{1.4}\times4 + 4\times e^{1.4}\times5 + e^{1.4}\times6]/3 \\ &= \$\ 3.230435871279753 \times 10^3 \approx \$\ 3230.44. \end{aligned}$$

The global absolute error is

$$e_1 = \$\ (6/90) \times 1^4 \times 1.4^4 \times e^{1.4\times6} = \$\ 1.138923441197582 \times 10^3 \approx \$\ 1138.92$$

and the relative error is

$$e_{1r} = e_1 \times 100/3230.44 = 35.25587845618554\% \approx 35.26\%.$$

This error is not sharp enough but certainly it shows that the computed result with h = 1 year has unacceptable error. Observe that the analytical integration provides the value of I as

$$I = \$\ 3.175761962642750 \times 10^3 \approx \$3175.76.$$

This implies that our choice of h is too crude (here too large). However, we do not have the knowledge of the analytical integration, in general. If we choose h = 0.3 and use the Simpson's 1/3 rule then

$$I = \$\ 3.176299643715818 \times 10^3 \approx \$\ 3176.30$$

while for h = 0.2 the rule gives

$$I = \$\ 3.175869403267828 \times 10^3 \approx \$\ 3175.87.$$

The global absolute errors are then

$$e_2 = \$\ (6/90) \times 0.3^4 \times 1.4^4 \times e^{1.4\times6} = \$\ 9.22527987370042 \approx \$\ 9.23$$

and

$$e_3 = \$\ (6/90) \times 0.2^4 \times 1.4^4 \times e^{1.4\times6} = \$\ 1.82227750591613 \approx \$\ 1.82,$$

respectively. The corresponding relative errors are

$$e_2 \times 100/3176.30 = 0.29058967981614\% \approx 0.29\%$$

and

$$e_3 \times 100/3175.87 = 0.05730713158914 \approx 0.06\%.$$

Evidently with smaller values of h, we get better quality of the results. This does not imply that for too small a value of h we get too good a result. This is because we have too much of computation for too small a value of h and hence non-negligible computational error which dominates. Also, with smaller h, computing time increases.

5.5.13 Numerical differentiation

The analytical differentiation of a function is a relatively easy and feasible task unlike the analytical integration which in most cases is not feasible. The numerical differentiation as well as integration on the other hand are easy and always feasible. However, sometimes analytical differentiation is undesirable since the derivative (e.g., derivatives with terms containing product of transcendental functions) might inflate physically as well as computationally and consequently would involve more computing resources and errors. In such cases, numerical differentiation has a scope in scientific computing. There are several numerical methods (Krishnamurthy and Sen 2001) to differentiate a function. We provide here a simple *extrapolatory* method called *the method of central difference limit* (Sen 2002; Krishnamurthy and Sen 2001) in which the successive step size is halved. The numerical derivative of the analytical function f(x) at $x = x_0$ is

$$df/dx = c_1[f((1 + 0.5\alpha)x_0) - f((1 - 0.5\alpha)x_0)]/(\alpha x_0) + c_2[f((1 + 0.25\alpha)x_0) - f((1 - 0.25\alpha)x_0)]/(0.5\alpha x_0) + c_3[f((1 + 0.125\alpha)x_0) - f((1 - 0.125\alpha)x_0)]/(0.25\alpha x_0),$$

where

$$c_1 = 1/3,\ c_2 = -2, \text{ and } c_3 = 8/3 \text{ (quadratic fitting).}$$

For some accuracy if we take $\alpha = 0.05$ then for the next higher order accuracy, we take $\alpha = 0.005$ or a value smaller than 0.005 (but not too small). However, to get an error-bound we must know the value of the derivative with higher-order accuracy and that with lower order accuracy. So, to obtain an error-bound here we compute df/dx at $x = x_0$ with two different α's – larger α corresponds to lower order accuracy while smaller α to higher order accuracy.

Consider the problem of computing the derivative of

$$f(x) = x^6 \text{ at } x = x_0 = 100.$$

If we choose $\alpha = 0.5$ then df/dx at $x = 100$ is

$$d_1 = 6.005126953124997 \times 10^{10} \text{ (lower order accuracy).}$$

If we choose $\alpha = 0.05$ then df/dx at $x = 100$ is

$$d_2 = 6.000000512695511 \times 10^{10} \quad \text{(higher order accuracy).}$$

Hence the relative error in d_1 in computing df/dx at x = 100 is

$$E_r = |d_2 - d_1|/|d_2| = |6.000000512695511 \times 10^{10} - 6.005126953124997 \times 10^{10}|/6.000000512695511 \times 10^{10} = 8.544066652392704 \times 10^{-4}$$

and the number of significant digits to which this df/dx, i.e., d_1 is correct is

$$\lfloor \log_{10}(1/E_r) \rfloor = \lfloor 3.06833537236451 \rfloor = 3.$$

Note that the exact derivative of x^6 at x = 100 is 6×10^{10} and also the derivative d_2 is more accurate than d_1. The foregoing relative error E_r is in d_1 and not in d_2. To compute the relative error in d_2 we need to compute a derivative, say, d_3 of still higher accuracy (e.g. taking α =0.005).

5.5.14 Ordinary differential equations–Fourth-order Runge-Kutta method

Any system of ordinary differential equations (ODEs) – linear or nonlinear, coupled or noncoupled – with requisite number of initial conditions can always be written as a system of n first order ODEs (Krishnamurthy and Sen 2001) with exactly n initial conditions. Consider the problem consisting of a system of n first order ODEs with n initial conditions (ICs):

$dy_i/dx = f_i(x, y_1, y_2, \ldots, y_n) \quad i = 1(1)n,$
ICs: at $x = x_0$, $y_i = y_{i0}$ $i = 1(1)n$.
Compute y_i $i = 1(1)n$ for $x = x_0 + h, x_0 + 2h, \ldots, x_0 + mh$. (h is specified)

The fourth-order Runge-Kutta scheme (Sen 2002; Krishnamurthy and Sen 2001) to solve this problem is as follows.
Compute

$k_{1i} = hf_i(x_0, y_{10}, y_{20}, \ldots, y_{n0}) \quad i = 1(1)n,$
$k_{2i} = hf_i(x_0 + h/2, y_{10} + k_{11}/2, y_{20} + k_{12}/2, \ldots, y_{n0} + k_{1n}/2) \quad i = 1(1)n,$
$k_{3i} = hf_i(x_0 + h/2, y_{10} + k_{21}/2, y_{20} + k_{22}/2, \ldots, y_{n0} + k_{2n}/2) \quad i = 1(1)n,$
$k_{4i} = hf_i(x_0 + h, y_{10} + k_{31}, y_{20} + k_{32}, \ldots, y_{n0} + k_{3n}) \quad i = 1(1)n$, and
$k'_i = (k_{1i} + 2k_{2i} + 2k_{3i} + k_{4i}) \quad i = 1(1)n.$
$y_{i1} = y_i(x_0 + h) = y_{i1} = y_{i0} + k'_i \quad i = 1(1)n.$

Error in this method is $O(h^5)$.

Having computed y_{i1} $i = 1(1)n$ replace y_{i0} by y_{i1} $i = 1(1)n$ in the foregoing steps to get $y_{i2} = y_i(x_0 + 2h)$ $i = 1(1)n$. Continue this process till we obtain $y_{im} = y_i(x_0 + mh)$.

The foregoing method could have been written in a more concise form by considering y, f, k_1, k_2, k_3, k_4, k' ***as n-dimensional vectors.***

Consider the ODEs

$dy_1/dx = 1/y_2$,
$dy_2/dx = x - 1/y_1$,
ICs: $y_1(0) = 1$, $y_2(0) = 1$.

Compute using the fourth order Runge-Kutta method, y_1, y_2 for x = 0.1(0.1)1. Also, obtain the error and accuracy in terms of the number of significant digits. Here h = 0.1, m = 10. Using the precision of 15 digits, we get:

x	y_1	y_2	
0.0	1.00000000000000	1.00000000000000	← ICs
0.1	1.10497995807272	0.90983645513787	
0.2	1.21963182950987	0.83867843908594	
0.3	1.34303768114535	0.78553836411816	
0.4	1.47359340254160	0.74945933647917	
0.5	1.60907125447432	0.72952973933015	
0.6	1.74681915406008	0.72490033390584	
0.7	1.88405840836056	0.73479851212980	
0.8	2.01820066684813	0.75853758267068	
0.9	2.14709973437236	0.79552012360594	
1.0	y_{1b}=2.26918793229602	y_{2b}=0.84523571855940	

To compute the error, viz., the relative error, we need the value of y_1 and that of y_2 with sufficiently higher order accuracy. For these values we re-compute y_1 and y_2 with smaller h and larger number of steps. Choose h = 0.01, m = 100 and compute y_1 and y_2 for x =0.01(0.01)1. We get

x	y_1	y_2	
0.00	1.00000000000000	1.00000000000000	← ICs
0.01	1.01004999811832	0.99009983395574	
0.02	1.02020001985863	0.98039870623502	
0.03	1.03044996760504	0.97089563286308	
0.04	1.04079963992723	0.96158957854416	
0.05	1.05124883539489	0.95247950798251	
0.06	1.06179735257770	0.94356438588236	
⋮			
0.98	2.24537126827786	0.83429483356558	
0.99	2.25731891860942	0.83970306049762	
1.00	y_{1g}=2.26918900117175	y_{2g}=0.84523465395560	

Let

$$\mathbf{y}_b = [y_{1b} \quad y_{2b}]^t, \quad \mathbf{y}_g = [y_{1g} \quad y_{2g}]^t$$

be the vectors representing the values of y_1 and y_2 at the last step (viz., x=1) of integration that has the maximum cumulative error in the step by step integration by the Runge-Kutta method. $\mathbf{y}_g$ is of sufficiently higher order accuracy and $\mathbf{y}_b$ is of lower order accuracy. The relative error and the accuracy in $\mathbf{y}_b$ are:

$$E_{rb} = ||\mathbf{y}_g - \mathbf{y}_b||/||\mathbf{y}_g|| = 6.230037933530361 \times 10^{-7},$$

Accuracy $= \lfloor \log_{10}(1/E_{rb}) \rfloor = \lfloor 6.20550930899528 \rfloor = 6$ significant digits.

Observe that the relative error and the accuracy in $\mathbf{y}_g$ are considerably less and considerably more, respectively. That is, $\mathbf{y}_g$ is certainly better than $\mathbf{y}_b$ in the quality of the result.

Most noniterative techniques do not provide the relative error as a part of the technique since result of sufficiently higher order accuracy needed for the error computation is often not available – one has to compute this result to know the error and hence the quality of the result. Observe that the step size h should not be taken too small so that the cumulative error due to excessive computation dominates significantly over the precision used.

5.6 Conclusions

In Section 5.5 we have presented only a few iterative and noniterative algorithms along with the related errors and complexity. We have numerous algorithms of different kinds. The error and complexity of each of them cannot be describes/presented in a single book nor are they required. The essence of error computation that pervades through the algorithms in Section 5.5 is more or less applicable in practically any algorithm. In fact, this essence is echoed in all the foregoing chapters on error. The knowledge of sufficiently higher order accurate quantity along with the present quantity (which is of lower order accuracy) constitutes the essence of numerical error computation in the present quantity. While dealing with matrices and vectors, from human psychology point of view, we have used matrix/vector norms to specify error in all concerned algorithms in Section 5.5. It may be observed that all the algorithms presented in Section 5.5 are polynomial time unless otherwise specified. One final point to note is that in a noniterative algorithm, the error computation could be twice or more (sometimes much more) as expensive as that for the solution itself.

Bibliography

Agrawal, M.; Kayal, N.; Saxena, N. (2002): *PRIMES is in P*, Unpublished manuscript, Indian Institute of Technology, Kanpur.

Abramowitz, M.; Stegun, A. (eds.) (1965): *Handbook of Mathematical Functions with Formulas, Graphs, and Mathematical Tables*, Dover Publications, Inc., New York.

Aho, A.V.; Hopcroft, J.E.; Ullman, J.D. (1974): *The Design and Analysis of Computer Algorithms*, Addison-Wesley, Reading, Massachusetts.

Alam, S.S.; Sen, S.K. (1996): *Computer and Computing in Fortran 77* (2nd Ed.), Oxford & IBH Publishing Co., New Delhi.

Fitzgerald, B.K.E. (1970): Error estimates for the solution of linear algebraic system, *J Res. Nat.Bur. Stand.,* **74B**, 251-310.

Gregory, R.T.; Krishnamurthy, E.V. (1984): *Methods and Applications of Error-free Computation,* Springer-Verlag, New York.

Karmarkar, N. (1984): A new polynomial-time algorithm in linear programming, *Combinatorics,***4**, 373-395.

Krishnamurthy, E. V.; Sen, S.K. (2001): *Numerical Algorithms: Computations in Science and Engineering*, Affiliated East West Press, New Delhi.

Lakshmikantham, V.; Sen, S.K.; Howell, G. (1996): Vectors versus matrices: p-inversion, cryptographic application, and vector implementation, *Neural, Parallel and Scientific Computations,***4**, 129-140.

Luenberger, D.G. (1973): *Introduction to Linear and Nonlinear Programming*, Addison-Wesley, Reading, Massachusetts.

Rokne, J.; Lancaster, P. (1971): Complex interval arithmetic, *Comm. ACM,* **14**, 111-112.

Sen, S.K. (2002): Error and Computational Complexity in Engineering, in *Computational Mathematics, Modelling and Algorithms* (Chap. 5) ed. J.C. Misra, Narosa Publishing House, New Delhi, 2002, 110-145.

Sen, S.K.; Prabhu, S.S. (1976): Optimal iterative schemes for computing Moore-Penrose matrix inverse, *Int. J. Systems Sci.*, **8**, 748-53.

Sen, S.K.; Ramful, A.; Jain, M.K. (2000): Solving linear differential equations as a minimum norm least squares problem with error-bounds *Intern. J. Computer Math*, **74**, 325-343..

Chapter 6

Error and Complexity in Error-Free, Parallel, and Probabilistic Computations

6.1 Introduction

6.1.1 Error in error-free computation?

For solving any numerical problem that involves a finite number of only basic arithmetic operations, viz., additions, subtractions, multiplications, and divisions, any error-free (exact) arithmetic that may or may not involve intermediate number growth could be used. The number of operations may or may not be known a priori. For ill-conditioned problems such as the solution of a near-singular linear system, quite often exact solution is required to avoid unacceptable error using the floating-point arithmetic. Any exact computation assumes that the input is exact. This assumption is in general untrue in the real world situation. Although the computation is error-less irrespective of the error-free algorithm used, the error in the input gets transmitted to the output (results) sometimes in a big way. Given the error-bounds for the input, we discuss the error-bounds of the output obtained through the exact computation. The error-bounds would be significant (large or broad) for specifically ill-conditioned problems even though the error-bounds in the input are small (narrow). It demonstrates that the problem of computing the error in the output is essentially an exponential algorithm even for linear problems. Consequently it provides some knowledge about the output error or, equivalently, an error estimate/bound for linear systems through an evolutionary/probabilistic approach which is always polynomial. We have also stressed that the error bounds for ill-conditioned (i.e., near-singular) linear systems could be sufficiently large. Consequently, the small uncertainty (i.e, small error bounds) in the input data due to, say random errors in measurement, could result in large error bounds. This implies that even the error-free computations for sensitive problems will have limited use in solving real world problems.

6.1.2 Error-free material universe/nature

The material universe/nature and all the natural computations that are continuously going on nonstop know no error (defined as the exact quantity minus the approximate quantity). The quantities in nature are real and are always exact and the results of natural computations are also always exact. It is humanly impossible to capture the exact quantities and the exact computations.

The material universe (MU) has always an absolute order in it. It is humanly impossible to unearth all the laws governing it (the order). We have unearthed some (not always exactly/precisely) but many are yet to be unearthed. There is no inconsistency (known to us) in this order. Observe that *the material universe is absolutely error-free.* Any thing/event that happens in it must follow the laws of nature exactly and, of course, sometimes some of these laws could be beyond our comprehension. Even the known laws may not correspond to the corresponding natural laws exactly. Thus the massively parallel process/computation that is going on in the universe/nature never had, has, and will have any error in the inputs, in the computations, and also in the outputs (results). No mathematical proof for all the foregoing statements exists. Nor anybody has disproved any of these statements. We assume here these assertions as experiential truths. Chaos meaning great disorder or confusion as well as the absolute randomness are two terms which are absent in the functioning of the universe/nature. We do, however, have specific definitions of chaos and randomness in mathematics (Baker and Gollub 1996). These definitions help us design polynomial-time algorithms to solve many real-world problems reasonably well while corresponding deterministic appproaches are exponential and hence intractable or too compute-intensive.

6.1.3 Erroneous domain of human activities

However, when we, the human beings, create a physical problem (PP) out of the MU, error creeps in the PP because of the inherently erroneous measuring device (to measure a quantity[1]) as well as due to assumptions made. It is also in-

---

[1]According to dictionary, a quantity is the "specified or definite amount or number; aspect of anything that can be measured, weighed, or counted". If the quantity is measured in terms of a nonnegative integer, e.g., 10 oranges, 15 mangoes, then it can be specified exactly. If the number (integer) is too large, for instance

troduced in all the subsequent stages, viz., mathematical model (MM), the algorithm (A), the high-level program (P_1) say, Matlab or Fortran program, intermediate program (P_2) such as the internal machine representation/reverse polish notation, . . ., the final machine language program (P_n) of a particular computer, and the computation (C). The required result (R) will be known to us in the form of a value which contains exact solution plus the cumulative error. Neither the exact solution nor the cumulative error is known to us explicitly. The foregoing stages, viz., PP, MM, A, P_1, P_2, . . ., P_n, are equivalent in the sense that each one is an ordered sequence of imperative sentences (commands). We have already seen in Figure 1.4 of Section 1.9 of Chapter 1 the introduction of errors in various stages while creating a physical problem from the material universe and then solving using a digital computer.

6.1.4 Error in hierarchical structure

A hierarchical structure may or may not be balanced. From a common human psychology point of view a balanced structure is desired. We have seen in Section 1.10 of Chapter 1 that, in a balanced hierarchical structure, if PP/MM has x_1 imperative sentences then A will have around $x_2 = (7 \pm 2)x_1$ imperatives sentences, P_1 will have around $x_3 = (7 \pm 2)x_2$ sentences and so on. The term (7 ± 2) is derived from the fact that a common human being can understand 7 ± 2 things at a time. However, in addition to the error in the input (PP), further errors are introduced into the stages MM, A, P_1, P_2, . . ., P_n, and C as these stages are not usually exactly identical in the sense that approximations creep in. Although some of these additional errors may be 0 sometimes, these errors are potentially there in a nonzero form.

The following diagram (Figure 6.1) which is a modified version of Figure 1.4 of Chapter 1 depicts the problem solving using a digital computer and the introduction of errors in various stages.

Result R is the exact solution (usually numerical) plus the cumulative error — none of these two is ever known in practice. Thus the result obtained out of a computer, though is certainly not the most desired exact solution, is a solution reasonably close to the exact solution, i.e., the error is reasonably low (non-dominant) compared to the magnitude of the solution provided the problem (MM) is well-posed and well-conditioned. It is, therefore, necessary to validate how good the result is. We will discuss this important issue in subsequent sections only in the context of error-free computation.

---

number of red blood cells in one mm^3 then the quantity in terms of the number is usually approximate (erroneous).

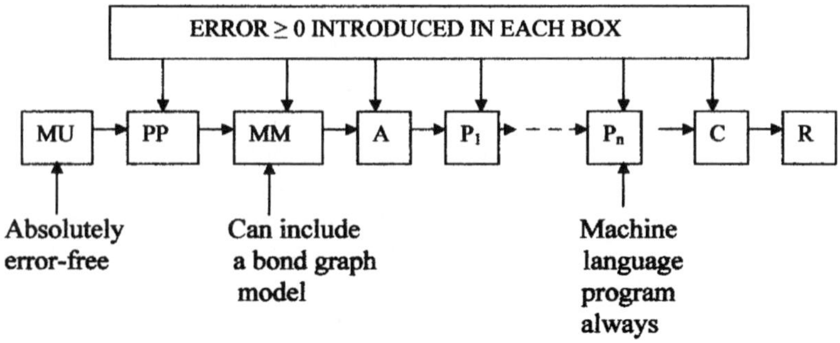

Figure 6.1: Schematic diagram of stages for problem-solving using a digital computer and error injection in various stages.

6.1.5 Bond graphs for describing dynamic models

However, we have already seen in Section 1.9 of Chapter 1 that for several continuous simulation problems, we may have immediately after the physical problem PP either a mathematical model MM or a bond graph model (Cellier 1998) or both in place of MM. The physical laws are made use of to derive a bond graph model for a given physical problem.

A bond graph depicts through a drawing the physical structure and the nature of subsystems. This graph can be derived from the physical model. The importance of the bond graph lies in the fact that when the physical model is partially changed, only the corresponding (and not the whole) part of a bond graph needs to be changed. For flow problems such as passive and active electric circuits current flow problems as well as for many mechanical systems, bond graph modelling is a convenient tool. The numerical solution of a bond graph model is often readily obtained using an appropriate available software exclusively meant for processing bond graphs with the specified inputs.

6.1.6 Why inconsistency in PP derived from consistent MU

Interestingly the PP derived from the error-free universe should not be inconsistent/near-consistent although sometimes it may be so because of (i) inherent nonremovable fixed order error ever existing in any measuring device as well as (ii) assumptions (if made). Such an inconsistent/nearconsistent PP is still solved in some sense (say, least-squares and/or minimum-norm sense) and the consequent result along with an estimate of error-bounds and complexity is almost always found useful in real-world applications. Even in this case error-free solution could be obtained along with some knowledge about the error. It may be observed that a perfectly consistent PP does not necessarily imply that it is error-free; in fact, quite often it is not.

6.1.7 Inconsistency in PP and possible error-free solution

Interestingly the PP derived from the error-free universe should not be inconsistent/near-consistent although it may be so because of (i) inherent nonremovable fixed order error ever existing in any measuring device as well as (ii) assumptions (if made). Such an inconsistent/nearconsistent PP is still solved in some sense (say, least-squares and/or minimum-norm sense) and the consequent result is almost always found useful in real-world applications. Even in this case error-free solution could be obtained along with some knowledge about the error.

6.1.8 Need for error-free computation

Error-free arithmetic, such as multiple-modulus residue arithmetic, p-adic arithmetic, rational arithmetic (practically not used because of *intermediate number growth*) could be employed only when the inputs are rational (ratio of two finite integers) and the number of operations in the algorithm used is finite (Gregory and Krishnamurthy 1984). For an *ill-conditioned* problem (a problem whose solution produced by using a finite-precision real/floating-point arithmetic has highly pronounced error) involving rational inputs, inexact arithmetic, viz., the real or the complex arithmetic gives totally erroneous results. This fact, usually known as the numerical instability shown in the foregoing section, has been demonstrated by many authors; see, or example, Kulisch and Miranker (1986).

6.1.9 Amplified error in error-free computation

To take care of such an unstable/sensitive situation, the superimposition of an error-free arithmetic on an algorithm is thus not only desirable but also a must in many ill-conditioned problems. Even in error-free implementation (which is possible when the algorithm/method involves only a finite number of arithmetic operations, viz., addition, subtraction, multiplication, and division), the inherent unavoidable error in the input gets magnified in the output results although the computation is 100% error-free (Sen and Mohanty 2003; Lakshmikantham et al. 2004).

6.1.10 The limitation of interval arithmetic

This error could be studied using *interval arithmetic* [Rokne and Lancaster 1971] as well as other methods. This study may be sometimes useful but several times more expensive both in terms of computing time as well as storage space. Also, the bounds of each exact real quantity may not be reasonably narrow (small) for the computation to be meaningful.

6.1.11 Why balanced hierarchical stages and error in each stage

We have already seen in Section 1.10 of Chapter 1 that starting from the physical problem PP to Computation C, there are n + 4 stages. If n = 3 (which is usually the case) then we will have 7 stages of problem-solving. The problem is created from the material universe MU and expressed as a PP which may not be corresponding exactly as the problem in MU and in all the n + 4 stages, errors are introduced, sometimes in some stages the error introduced, however, could be zero. An interesting aspect of the step-by-step problem-solving is the equivalence and hierarchical structure of the 7 stages (from PP to CP_n both inclusive for n = 3). These 7 stages are equivalent in the sense that each of these stages is a set of imperative sentences (commands/instructions to the computer) along with input data which are usually assertive sentences or information needed for processing. We measure the size of each stage as the number of imperative sentences/instructions it has. If the PP has k imperative sentences then the mathematical model MM has around k imperatives sentences. For a *balanced hierarchical structure*, the algorithm A should have $k_1 = (7 \pm 2)k$ instructions; the high-level programming language CP_1 , say Fortran, will have $k_2 = (7 \pm 2)k_1$ instructions, while the internal machine representation (not visible by the user) CP_2 should have $k_3 = (7 \pm 2)k_2$ instructions and the machine language program CP_3 (when n = 3) should have $k_4 = (7 \pm 2)k_3$ instructions. Thus we see that every succeeding stage starting from MM to CP_n will be around 5 to 9 times larger physically. If k = 4 and if each succeeding stage is 7 times larger then the corresponding machine program will have $4 \times 7^4 = 9604$ instructions. A set of some of these instructions will be usually executed by the computer repeatedly/iteratively thus the actual number of instructions to be executed by the computer will be usually much more than 9604. The factor 7 ± 2 is introduced since psychologically a common human being can grasp/remember 7 ± 2 things at a time. He can certainly not remember too many items, say 1000 names, at a time when told to him once. It is, however, possible that there exists a superhuman/uncommon human being who could remember very large sequence of names.

6.1.12 Man-man, man-machine, and machine-machine communications

The foregoing limitation of a common human being, viz., not being able to remember beyond 5 to 9 items at a time, is important in *man-man* (including man-self) communication as well as *man-machine* communication. So far as *machine-machine* communication is concerned, there is no such limitation. The nonliving machine can remember millions of things, say names, for an indefinite period of time exactly once these are given to it subject, however, to its memory capacity. The communication from one machine to another knows no such limitation. As such we may certainly write unbalanced hierarchical structure of a stage (algorithm/program). The machine will face absolutely no problem and it would produce the desired result. Man, however, could have difficulty in grasping as well as in modifying/debugging. Further, while an unstructured program is bad for man-man communication, it produces no problem for the man-to-machine as well as machine-to-machine communication.

We show, in Section 6.2, that determining an error-bound in the output of an error-free computation for a solution vector **x** of the given equation A**x** = **b** is an exponential problem and hence intractable. Also it includes a probabilistic approach which is polynomial-time to obtain an estimate of error in the output. In addition it comprises illustrative numerical examples and some remarks.

In Section 6.3, we demonstrate that the quality of numerical output in a parallel computation is at least as good or better than that in sequential computation. We mention further, in this section, that the order of parallel complexity is usually lower than the corresponding sequential complexity by one.

The error-bound in the probabilistic algorithm called PAC learning model is discussed in Section 6.4 and is shown to be valid only under a confidence level which is never 100%. Yet another probabilistic algorithm based on the shrinking rectangle algorithm is presented in Section 6.5. The probabilistic algorithms are always polynomial-time and are often used for intractable (exponential) problems, specifically optimization problems which may not have polynomial-time algorithms. A polynomial-time (probabilistic) algorithm devised for exponential problems will always produce results on which we can never place 100% confidence unless the verification of the results is polynomial-time.

6.2 Actual error-bound in exact computation: Exponential problem

Consider the linear system A**x** = **b**, where A is an m × n given numerical matrix and **b** is an m-column vector. The erroneous inputs are the matrix A and the column vector **b**. Since a measuring device has always associated with it a fixed order of error, it will not be wrong to say that each element of A and that of **b** measured by a single device has the order of error 0.05% (say). The order of error will not differ from element to element. A screw gauge, for instance, to measure the diameter of the order of 10 cm of a sphere would have an error ±0.5 mm, i.e. 0.05% (say). This error (rather error-bound) is fixed for this screw gauge. This information tells us the precise bound of each element of A and that of **b**. For example, the exact value of the element a_{ij} will lie in the interval $[a_{ij} - .0005 \times a_{ij}, a_{ij} + .0005 \times a_{ij}]$. So is the case with b_i. Over a small interval for an element of A, the error will behave linearly[2]. For m × n elements of the matrix

[2]If the elements of A are exact and those of **b** are in error, then the error in the solution vector of A**x** = **b** will be linear. We have $A\mathbf{x}' = \mathbf{b} + \Delta\mathbf{b}$. Then $\mathbf{x}' = A^{+}\mathbf{b} +$

A, we have m × n lower bounds,, viz., $a_{ij} - .0005 \times a_{ij}$ and m × n upper bounds, viz., $a_{ij} + .0005 \times a_{ij}$. Similarly, for b, we have m lower bounds $b_i - .0005 \times b_i$ and m upper bounds $b_i + .0005 \times b_i$. Thus we have mn + m lower bounds and mn + m upper bounds for the inputs A and **b**. Hence we will have to consider 2^{mn+m} possible combinations of bounds and solve 2^{mn+m} linear systems exactly. These solutions will produce the lower and upper bounds of each element of the solution vector. Unfortunately, the procedure is exponential, i.e., the computational complexity will be $O(mn^2) \times e^{(mn+m)\log 2}$. Hence this procedure is inefficient (slow) and is intractable (though Turing (Turing 1936) solvable). If m = n = 10 and if the computer executes 10^8 operations/sec then the time complexity only to obtain the error bounds will be $(mn^2 \times e^{(mn+m)\log 2})/(10^8 \times 60 \times 60 \times 24 \times 365)$ years = 4.1162×10^{20} years which is larger than the estimated age of the universe! For a matrix of order 7, the time complexity only to obtain the error bounds is 7837.3 years. The foregoing assumption of small (narrow) bounds on the elements of A, that permits us to consider the error to be linear may not be all that true for ill-conditioned, i.e., sensitive (e.g., near-singular) systems. Under these circumstances, we can get an error estimate using an evolutionary approach. This will be discussed in the next sub-section.

6.2.1 Evolutionary approach for error estimate in exact computation

Consider the m × (n + 1) augmented matrix $D = [d_{ij}] = [A, \mathbf{b}]$ of the system $A\mathbf{x} = \mathbf{b}$. Let the error introduced in each element of D be 0.05% and D′ be the m × (n + 1) matrix whose each element d'_{ij} is an ordered pair $[d_{ij} - 0.0005d_{ij}, d_{ij} + 0.0005d_{ij}]$. The evolutionary procedure is as follows.

S. 1 Compute error-free the solution $\mathbf{x}_c$ of the system represented by D.
S.2 Generate m × (n + 1) uniformly distributed random (pseudo-random) numbers in the interval $[d_{ij} - 0.0005d_{ij}, d_{ij} + 0.0005d_{ij}]$. Call the resulting augmented matrix D_r.
S.3 Compute error-free the solution $\mathbf{x}_r$ of the linear system represented by D_r.
S. 4 Obtain the relative error (in the solution vector $\mathbf{x}_r$) $e_k = ||\mathbf{x}_c - \mathbf{x}_r||/||\mathbf{x}_c||$.
S. 5 Repeat S.2-4 for k = 1 to s (=100, say) times.
S. 6 Obtain the largest e_k — this will give an estimate of relative error-bound for the error-free computation.

$A^+\Delta\mathbf{b}$, where A^+ is the minimum norm least squares inverse of A The error term $A^+\Delta\mathbf{b}$ is linear in $\Delta\mathbf{b}$. The error term will not be linear if the elements of A are not exact. Assume **b** is exact; A and A + ΔA are non-singular. The solution of the system $(A + \Delta A)\mathbf{x}' = \mathbf{b}$ can be written as $\mathbf{x}' = (I + A^{-1}\Delta A)^{-1}A^{-1}\mathbf{b} = [I - A^{-1}\Delta A + (A^{-1}\Delta A)^2 - \ldots]A^{-1}\mathbf{b}$. If the error bounds in the elements of A are small, i.e., ΔA is small then we neglect second and higher order terms. The error thus would behave linearly for small bounds.

This probabilistic approach is polynomial time $O(smn^2)$, where s is independent of m and n. We may compute the mean and standard deviation of the s errors e_k. These computations will reveal the degree of sensitivity (ill-conditioning) of the system. By varying the value of s and computing the corresponding standard deviation, we would get a relationship of the standard deviation against s. This relationship will give an estimate of our confidence that we can place on the error-free solution.

Confidence estimate For different values of s (= uniformly distributed random numbers), say, 100, 150, and 200, we compute the standard deviations d_i of the errors. To be 100% confident will be impossible. However, the difference between successive standard deviations for increasing s should be usually a monotonically decreasing (more strictly nonincreasing) function of s. Once this happens, we compute the confidence estimate as ((1 – the difference of last two successive standard deviations)/(last standard deviation)) × 100.

6.2.2 Numerical experiment

As mentioned earlier that the error-free computation is usually prescribed for ill-conditioned (highly sensitive) linear system. Exactly singular linear systems do not pose problems. We may remove exactly linearly dependent rows of the consistent system as these rows are redundant (Sen and Sen 2004). The resulting shrunk system with full row rank can then be solved with good accuracy as long as the shrunk system did not have nearly linearly dependent rows. The real problem crops up for near-singular system just as we face real problem for obtaining roots of a polynomial having root-clusters (closely-spaced roots). Observe that exactly repeated roots do not pose problem with most algorithms.

An ideal set of matrices for testing the effect of slight perturbation of the inputs on the output solution is that of Hilbert matrices. The higher the order of the Hilbert matrix is, the more pronounced is the near-singularity. Hence we consider Hilbert matrices to demonstrate that error-free computation of a solution vector $\mathbf{x}$ of the system $A\mathbf{x} = \mathbf{b}$ will have extremely highly magnified error for a very small perturbation of the input matrix A and vector $\mathbf{b}$. Most often the slight variation in A and $\mathbf{b}$ is due to uncertainty/fixed-order error in measuring devices.

Consider the 8 × 8 Hilbert matrix A whose (i, j)th element is $1/(i + j - 1)$ and the vector $\mathbf{b}$ whose i-th element is the sum of the elements of the i-th row of A. The exact solution vector $\mathbf{x}_e$ will then be $[1\ \ 1\ \ 1\ \ 1\ \ 1\ \ 1\ \ 1\ \ 1]^t$. The augmented matrix is $D = [A, \mathbf{b}]$.

Now we consider the random augmented matrix D_r where each element of D_r is the corresponding element of D multiplied by a random number distributed uniformly between $(1 - \varepsilon)$ and $(1 + \varepsilon)$. If we take $\varepsilon = 10^{-10}$ which is too small compared to the uncertainty/error in a measuring device then the relative error in the output would be as high as 10%. In practice ε and hence the output error will be much greater.

The linear system D_r was solved to yield the solution $\mathbf{x}_r$. The maximum error and confidence estimates were then obtained using the methods of the next subsection. All calculations were done using the program "GP/PARI" with 50 digits of precision which is assumed equivalent to error-free computation by any exact arithmetic such as the multiple modulus residue arithmetic, p-adic arithmetic, rational arithmetic for small problems like the ones involving low order Hilbert matrices. Table 6.1 provides error and confidence estimates for the 8 × 8 Hilbert system.

Table 6.1 Relative error and confidence estimates of the extremely small perturbed solution of 8 × 8 Hilbert system

Number of Trials	Maximum relative error (e_k)	Confidence (%)
100	0.08706	------
200	0.10109	94.174
300	0.10109	95.596
400	0.10109	98.899
500	0.10109	99.891
600	0.10109	98.016
700	0.10109	98.953
800	0.10109	99.576
900	0.10109	99.691
1000	0.10109	99.169

Consider the system $A\mathbf{x} = \mathbf{b}$, where A = the 5 × 5 Hilbert matrix and $\mathbf{b}$ is the vector of row-sums as before so that the solution $\mathbf{x}_e = \mathbf{e} = [1 \quad 1 \quad 1 \quad 1 \quad 1]^t$. Let $D = [A, \mathbf{b}]$. If we now make the (1, 2)nd element of A as $0.5 - 0.0005 \times 0.5$, (2,3)rd element as $0.25 + 0.25 \times 0.0005$, 1st element b_1 of $\mathbf{b}$ as $b_1 - 0.0005 \times b_1$ and the 3rd element as $b_3 + 0.0005 \times b_3$, then we obtain the solution (retaining up to 5 decimal places) $\mathbf{x}_r = [0.34667 \quad 10.45297 \quad -33.70323 \quad 47.92265 \quad -20.17300]^t$ which has the relative error $\|\mathbf{x}_e - \mathbf{x}_r\|/\|\mathbf{x}_e\| = 28.086$. This error implies that the perturbed solution is over 2800% off the exact solution $\mathbf{e}$ even for a very small Hilbert matrix. This is just to show that error-free computation has a limited scope in the real world applications.

6.2.3 Remarks

Scope of error-free computations Error-free computations in most real world problems where these could be used are not attempted. This is because the floating-point arithmetic with the available precisions is good enough. However, for truly bad behaved problems, i.e., problems which produce very large change in their outputs for a small change in the inputs (for a specified precision), error-free computations are recommended. The bad behaved problems will, however, be good behaved if we allow sufficient larger precision (word-length). In most digital computers, the precision cannot be arbitrarily increased for floating-point computations. Thus, error-free computation is a reasonable alternative to obviate this problem of the limitation of precision. If the problem is too bad behaved but solvable, then certainly exact computation would be a must. It is, however, necessary in many real world situation to determine the effect of a small perturbation of the initial data on the output results so that we would be able to get to know the behaviour of the problem better. Our task would not be complete until we have studied the perturbation effect on the result.

Confidence level in the error estimate of exact solution With which confidence can we accept the estimated error-bounds? This question has been discussed in Sections 6.3 and 6.4. Although we could not obtain a precise confidence level for the error estimate, we did obtain an estimate of confidence through computing the standard deviations against different values of the parameter s — the number of random numbers generated.

Errorfree computation in linear optimisation problems A linear optimisation problem or, equivalently, a linear program had been an NP problem until 1979. Khachian's ellipsoid method (Khachian 1979) was the first polynomial method in only integer model (Traub and Wozniakowski 1982). Karmarkar's projective transformation method (Karmarkar 1984) was the first polynomial method in real number model. Error-free arithmetic cannot be implemented in both these methods to solve a linear program. However, the most widely used simplex method as well as several other methods (Lakshmikantham et al. 1995, 1996, 1997, 2000; Sen and Maulloo 1994; Sen and Howell 1992; Sen 2003) are amenable to error-free computations although these are iterative. In fact, error-free arithmetic can be superimposed on almost all deterministic iterative methods for linear programs. The exponential (combinatorial) method (for linear programs with finite optimal solution) based on Fundamental Theorem of Linear Programming, though intractable, is amenable to exact computation.

6.3. Parallel computation: error and complexity

6.3.1 Lower order error-bound in parallel summation

While no general claim can made about the difference in quality of results in serial as well as parallel computations, we can show that the outputs of some parallel computations do have lower order error-bounds. Consider, for example, the sum

$$S_N = \Sigma\, a_j,$$

where the summation is over $j = 1$ to $N = 2^n$.

Serial Mode In the serial mode, we may write

$$S_0 = 0,\ S_j = S_{j-1} + a_j,\quad j = 1, 2, \ldots, N.$$

Because of rounding errors, we actually deal with approximate values S'_j and a'_j. Thus

$$S'_0 = 0,\ S'_j = S'_{j-1} + a_j + r_j\ ,$$

where $r_j = a'_j - a_j$. Let

$$e_j = S'_j - S_j = \Sigma\, r_k,$$

where the summation is over k = 1 to j. In the floating point representation in a digital computer, if the mantissa has s binary digits, then S'_j can be written as

$$|S'_j| = 0.b_1b_2b_3 \ldots b_s \times 2^{p(j)}.$$

If we allow $a = \max |a_j|$, $j = 1, 2, 3, \ldots, N$, then

$$|S_j| \le ja,\ 2^{p(j)} \le 2ja.$$

Hence, allowing the following summation over i = 1 to N, we have

$$|e^N| = |S'_N - S_N| \le \Sigma\, |r_i| < \Sigma\, 2^{-s}2^{p(i)} \le 2^{-s+1}a\ \Sigma\, i = 2^{-s}aN(N + 1),$$

Thus the error in serial mode is $O(\alpha N^2)$, where the constant $\alpha = 2^{-s}a$.

Parallel Mode Let

$$S_{0j} = a_j,\ j = 1, 2, 3, \ldots, N,$$

$$S_{kj} = S_{k-1,2j-1} + S_{k-1,2j},\quad k = 1, 2, 3, \ldots, \log_2 N,\ \ j = 1, 2, 3, \ldots, N/2^k.$$

Proceeding as in the foregoing serial mode, we obtain

$$|S_{kj}| \le 2^k a,\ 2^{p(k)} \le 2 \times 2^k a.$$

Hence the error at each computation of S_{kj} is e_{kj}, where

$$|e_{kj}| \le 2^{-s} \times 2 \times 2^k a.$$

Let $r = N/2^k$. We obtain, for the r-th partial sum of the k-th step, for the error $e_k = \Sigma\, e_{kj}$, where the summation is over j = 1 to r, the bound

$$|e_k| \le r \times 2^{-s} \times 2 \times 2^k a = Na \times 2^{-s+1}.$$

Hence the total error $e = \Sigma e_k$ has the bound

$$|e| \le \Sigma\, Na \times 2^{-s+1} = Na \times 2^{-s+1} \log_2 N = O(\beta N \log_2 N),$$

where the summation is over $k = 1$ to $\log_2 N$ and the constant $\beta = 2^{-s+1}a$.

This error bound $O(\beta N \log_2 N)$ in the parallel mode is an improvement over the error bound $O(\alpha N^2)$ derived in the serial/sequential mode. That is, the error-bound in parallel mode is narrower (sharper) than that in serial mode. This, however, does not imply that the quality of actual output will be better in parallel computation than in sequential computation.

The computational complexity of serial summation is N – 1, i.e., O(N) while that of parallel summation is $O(\log_2 N)$. This is because only one processor is used in serial summation while N/2, N/4, N/8, . . ., 1 processors are used in parallel summation, N being the number of elements to be summed up. If $N = 2^{17}$ =131072 , then the parallel summation will need 17 units of parallel time, while serial summation will require 131071 units of (serial) time (neglecting communication overheads although these may not be all that negligible). The foregoing summation is one of the simplest examples.

6.3.2 Parallel algorithms: Efficiency and complexity

An algorithm is essentially a precise description of how to carry out a task or process. The efficiency of an algorithm is expressed in terms of the problem size, also called the input size, for which it is designed. The problem size is the length of the input of the problem, i.e., the number of symbols 0 and 1 that are necessary to write the length in the binary number system. For example, consider the inequalities

$$\Sigma\, a_{ij}x_j \le b_i, \quad i = 1, 2, 3, \ldots, m$$

where the summation is over j = 1 to n, x_j are real, a_{ij}, b_i are integer, $m \ge 2$, and $n \ge 2$. Then the problem size is

$$L = (\ \Sigma\Sigma \log_2 (|a_{ij}| + 1) + \Sigma \log_2 (|b_i| + 1) + \log_2 mn) + 1,$$

where the double summation is over i = 1 to m and j = 1 to n while the single summation is over i = 1 to m.

We now define polynomial number of processors (used to solve a problem of size L) which is a polynomial in L. In practice, at most a polynomial number of processors is reckoned to be feasible. In a sequential algorithm, the instructions are executed one after the other in sequence by a sequential computer, viz., the von Neumann computer. In a parallel algorithm, the instructions are executed simultaneously in some sense and are carried out by a parallel computer. A parallel algorithm is called a ***polylogarithmic*** time algorithm when it takes $O(\log^k$

L) parallel time for some constant integer k, where L is the problem size. The algorithm is universally regarded as efficient if it a polylogarithmic time algorithm using a polynomial number of processors. These algorithms are said to belong to ***class NC*** (Nick-Pippenger's class). An optimal parallel algorithm is an algorithm for which the product of the parallel time t and the number of processors p used is linear in the problem size L, i.e., pt =O(L). An optimal parallel algorithm for a problem may also mean that algorithm for which the product pt is equal to the computation time of the fastest known sequential algorithm for the problem. We have seen in Section 3.2 of Chapter 3 that algorithms whose asymptotic behavior is 2^n or c^n for some positive constant c are exponential algorithms or slow algorithms. These algorithms are infeasible for all but extremely small input data sizes. On the other hand, algorithms whose asymptotic behaviour is n, n^2, or more generally, n^c for some positive constant c are polynomial time algorithms or, simply polynomial algorithms or fast algorithms. Polynomial algorithms are usually feasible for practical input data sizes.

For any problem (such as, any NP-complete problem, see Sec. 2 of Chap. 3) ***for which there exists no known polynomial time sequential solution, a fast (efficient) parallel solution using a polynomial number of processors cannot be expected to be found.*** However, such a parallel solution for a problem with polynomial time sequential algorithm (i.e., a problem in ***class P***) may be desired. Many such problems do not seem to be readily parallelizable. These problems which are hardly parallelizable form the class of so-called P-complete problems. If an efficient parallel solution for any P-complete problem could be obtained then a similar solution would be found for any other. It is believed (not proved) P ≠ NC. See also Chapter 3.

The model presently used to describe numerical algorithms in the context of digital control-flow computers is the universal model of von Neumann, also known as serial universal model (Figure 6.2).

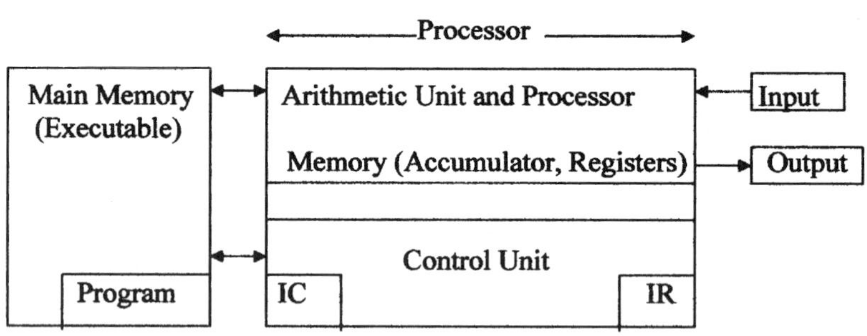

Figure 6.2: Universal serial model (computer)

In this model, IC denotes the instruction counter containing the address of the current instruction being executed, while IR represents the instruction register that contains a copy of the current instruction. The features of the universal computer are (a) digital (binary) representation of variables and constants, (b) sequential processing, i.e., executing one instruction at a time, and (c) holding the program, i.e., storing the coded version of the algorithm and data in the main memory.

The serial computer has improved enormously during the last fifty years due to technological and architectural innovations. The introduction of cash memory has increased the processing speed of serial machine. Some degree of parallelism has been introduced. This includes (a) overlapping of input/output (IO) instructions and non-IO instructions using channels (IO processors) and other means and (b) interleaved main memory techniques. This degree of parallelism soon fell below the requirements of numerical methods even at the programming level.

In 1960's the possibility of constructing truly parallel digital computers began to be explored. Soon models of parallel computers were designed and developed. A need for a ***parallel numerical analysis***, i.e., a body of numerical mathematics that takes specific advantage of the possibilities offered by parallel computers exists.

In parallel numerical analysis, the speed gain expected from the simultaneous/overlapped operation of p processors is important. So, an operational measure S, called the ***speed-up ratio***, is used. It is defined as follows. Let T_1 = computing time using on a serial machine (i.e., one-processor machine) and T_p = computing time on a parallel machine (i.e., p-processor machine with $p \geq 1$). Then the speed-up ratio is

$$S_p = T_1/T_p \geq 1,$$

where for a given problem the best available serial algorithm and the best available parallel algorithm are used even if these are different. Let $0 < k < 1$ be a machine dependent quantity and p be the number of processors. Then S has four forms, viz.,

(a) $S_p = k$ (for some nonlinear recursions),
(b) $S_p = kp$ (for matrix computations),
(c) $S_p = k \log_2 p$ (for searching), and
(d) $S_p = kp/\log_2 p$ (for linear recursions, sorting).

In order to assess the merits of a parallel algorithm, the parallel machine will be assumed to satisfy the following constraints.

(a) Given a finite number of processors, all processors are available at every moment,
(b) Each processor can carry out each of the four basic arithmetic operations in a single time unit. Further, in a specified time step, all processors execute the same instruction, i.e., the machine is SIMD (Single Instruction Multiple Data), and
(c) No limitation exists on data and storage.

The efficiency of utilization of the parallel machine is

$$E_p = S_p/p \le 1.$$

The cost of the algorithm is

$$C_p = pT_p.$$

The effectiveness to compare two parallel algorithms for the same problem is

$$F_p = S_p/C_p.$$

In order to illustrate these terms, let us consider the summation Σ a_i, where the summation is over i = 1 to 32. The following table (Table 6.2) of performance measures can then be easily checked for various number of processors p:

Table 6.2 Performance measures

p	T_p	C_p	S_p	E_p	F_p
1	31	31	1	1	.03
2	16	32	1.94	.97	.06
3	13	39	2.38	.79	.06
4	9	36	3.44	.86	.10
8	6	48	5.17	.65	.11
16	5	80	6.20	.39	.08

Observe that the effectiveness F_p is a measure of both the speed-up and efficiency. A parallel algorithm is considered effective if it maximizes F_p. From the foregoing table, it can be seen that efficiency E_p decreases and S_p increases steadily as the number of processors p increases while F_p is maximum when p = 8. Thus, the best choice of the number of processors for the foregoing summation is p = 8.

A general configuration of a parallel computer, though too general to precisely state its functioning, is given in Figure 6.3.

The parallelism may exist within the control unit, among the processors P_i and stores S_j, and in the central networks.

The processor organization in parallel computers may be through a mesh, a pyramid, a shuffle-exchange, a butterfly, a hypercube, or a cube-connected cycles network (Quinn 1987).

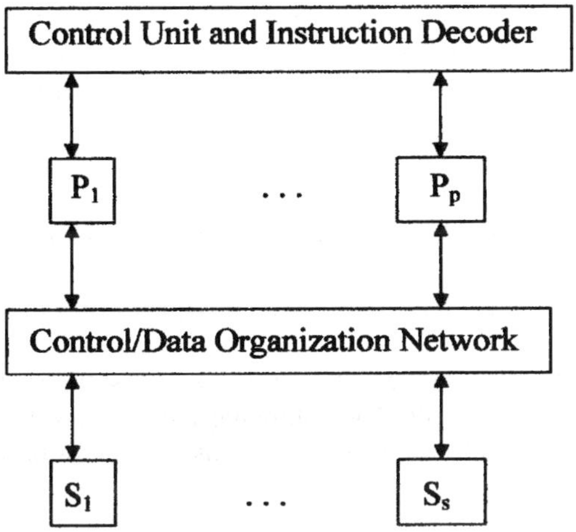

Figure 6.3: General model of a parallel computer

It has been shown that a universal parallel computer can simulate every reasonable parallel model with only a small loss of time and with essentially the same number of processors. The heart of this universal computer is a sorting network that is used as a post office for sending and requesting information. It has also been shown that since the cube-connected cycles are used as the sorting network, the cube-connected cycles network (and hence the butterfly) is an efficient general-purpose network. For further information, the reader may refer Quinn (1987).

A principle for the construction of a parallel algorithm for most SIMD (Single Instruction Multiple Data) computers is to start with a serial algorithm and convert it into a procedure which operates on ***vectors/matrices*** since vector/matrix operations can be carried out in parallel.

Another principle, a very general one, is the method of ***recursive doubling***. Consider the set of $n = 2^m$ elements e_i, $i = 1, 2, \ldots, n$. Let op (for example, op $\in \{+, *, \max, \ldots\}$) be an arbitrary associative operation that can be carried out on the set. Now the expression e_1 op e_2 op $e_3 \ldots$ op e_n needs $\log_2 n$ parallel steps for its evaluation against $n - 1$ serial steps. The tree structure appears to be a basic structure in parallel computations and this may occur in many implicit forms. Hence one may attempt to discover a tree structure underlying a given serial algorithm. As an example, consider the problem of finding the maximum of $n = 2^m$ elements, where m is a positive integer. If m is not a positive integer then a minimum number of dummy elements can always be added to ensure that n is a power of 2.

In the first step, maximum of each of the n/2 pairs of elements is found using n/2 processors working in parallel (simultaneously). In the second step, maximum of each of the n/4 pairs of resulting maxima is determined using n/4 processors and so on.

A balanced binary tree is the basic structure of this method. The balanced binary tree method is sometimes termed as ***log-sum algorithm***. To achieve log-sum algorithm, an expression E_0 should be such that

$E_0 = f(E_1E_2)$, where f is a function,
E_1 and E_2 are independent processes and are of equal computational complexity, and
Computation of E_1 and E_2 needs the same sequence of operations.

Decomposing E_1 and E_2 further according to the foregoing principles, we arrive finally by the principle of recursive doubling at E_0. ***The operations which are not associative are not capable of resolution by this fast and direct algorithm.***

The depth of the tree is m and the complexity of such an algorithm is $O(\log_2 n)$. It may be observed that the binary tree method is a bottom-up process. Yet another example is the ***second order recurrence formula which is essentially a serial process***:

$$f_0 = 1,\ f_1 = b_1,\ f_i = b_i f_{i-1} + a_i f_{i-2},\quad i = 2, 3, 4, \ldots, n.$$

This formula can be written using the matrix vector notation as follows. Let

$$q_j = \begin{bmatrix} f_j \\ f_{j-1} \end{bmatrix} \quad R_i = \begin{bmatrix} b_i & a_i \\ 1 & 0 \end{bmatrix},\ j=1, 2, 3, \ldots, n,\ i=2, 3, 4, \ldots, n.$$

Then

$$q_i = R_i q_{i-1},\quad i = 2, 3, 4, \ldots, n.$$

Since the matrix vector multiplication is associative, q_n can be computed using the recursive doubling technique in $O(\log_2 n)$ steps against $O(n)$ steps using a sequential algorithm.

The third principle is the ***divide and conquer*** technique. A given problem is divided into a number of subproblems which are recursively solved subject to the condition that the solution of the problem at any level of recursion are constructible from the solutions of its subproblems. The parallel computation time for the problem is proportional to the depth of recursion. In parallel computation, this technique needs that the subproblems at the same level of recursion should be independently solved. The foregoing binary tree method may be viewed as a special case of divide-and-conquer technique but from the point of view of the tree structure this technique is a top-down process. As an example, consider the evaluation of the polynomial $p_n(x)$ of degree $n = 2^m - 1$, where m is a positive integer, at $x = x_0$. Write

$$p_n(x) = x^{(n+1)/2}q(x) + r(x),$$

where q(x) and r(x) are polynomials of degree $2^{m-1} - 1$. Compute $q(x_0)$ and $r(x_0)$ recursively in parallel and combine the results to get $p_n(x_0)$. Th depth of recursion is $O(\log_2 n)$ here.

As a specific example, consider the polynomial $\Sigma\, a_i x^i$, where the summation is over i = 0 to 7. Write the polynomial as $r(x) + x^4 q(x)$, where

$$r(x) = a_0 + a_1x + a_2x^2 + a_3x^3,\ q(x) = a_4 + a_5x + a_6x^2 + a_7x^3.$$

In the next level of recursion, write $r(x) = r_1(x) + x^2q_1(x)$, where

$$r_1(x) = a_0 + a_1x,\ \ q_1(x) = a_2 + a_3x$$

and $q(x) = r_2(x) + x^2q_2(x)$, where

$$r_2(x) = a_4 + a_5x,\ q_2(x) = a_6 + a_7x.$$

In the same way write the next, i.e., the final level of recursion. The numerical computation then starts from this final level and each level is computed in parallel.

Another technique known as ***collapsing/compression technique***, may be illustrated by considering the summation of $n = 2^m$, where m is a positive integer, elements e_i stored in an array. Recursively, for each alternate value of i in parallel, compress the two entries e_i and e_{i+1} into a single entry with the value of the result of addition $e_i + e_{i+1}$. The length of the array is halved from one level of recursion to the next. Thus the array finally results to a single element after a logarithmic number of levels. This technique is essentially the same as the balanced binary tree method. In fact, all the foregoing techniques are not totally disjoint.

6.3.3 Numerical problems: Parallel implementation

We now consider a few numerical problems and the parallel implementation of their solution.

Linear recurrence scheme R<n, m> A linear recurrence scheme R<n, m> of order m for n equations, where m < n, is defined as

$$R\langle n,m\rangle:\ x_k = \begin{cases} 0, & k \le 0 \\ c_k + \Sigma a_{kj}x_j, & k=1, 2, 3, \ldots, n \end{cases},$$

where the summation is over $j = k - m$ to $k - 1$. If $m = n - 1$, then this system, denoted by $R<n>$, is known as ***ordinary linear system of recurrence equations***. The system $R<n, m>$ can be written in matrix-vector notation as $\mathbf{x} = \mathbf{c} + A\mathbf{x}$, where $\mathbf{x} = [x_1 \;\; x_2 \;\; x_3 \ldots x_n]^t$, $\mathbf{c} = [c_1 \;\; c_2 \;\; c_3 \ldots c_n]^t$, A is strictly lower triangular matrix of appropriate order with $a_{ik} = 0$ for $i \le k$ or $i - k > m$. A is a band matrix for $m < n - 1$. The system $R<4, 2>$ can be explicitly written as

$$
\begin{aligned}
x_1 &= c_1,\\
x_2 &= c_2 + a_{21}x_1,\\
x_3 &= c_3 + a_{31}x_1 + a_{32}x_2,\\
x_4 &= c_4 \qquad\quad\;\; + a_{42}x_2 + a_{43}x_3.
\end{aligned}
$$

The system $R<5>$ can be explicitly written as

$$
\begin{aligned}
x_1 &= c_1,\\
x_2 &= c_2 + a_{21}x_1,\\
x_3 &= c_3 + a_{31}x_1 + a_{32}x_2,\\
x_4 &= c_4 + a_{41}x_1 + a_{42}x_2 + a_{43}x_3,\\
x_5 &= c_5 + a_{51}x_1 + a_{52}x_2 + a_{53}x_3 + a_{54}x_4.
\end{aligned}
$$

To solve the foregoing $R<n>$ system with O(n) processors in O(n) time unit steps a method known as the ***column-sweep algorithm*** may be used. The algorithm can be described as follows.

- • x_1 is known. Compute $c_i^{(1)} = a_{i1}x_1 + c_i$ in parallel. Now x_2 is known and we have $R<n - 1>$ system.
- • x_2 is known. Compute $c_i^{(2)} = a_{i2}x_2 + c_i^{(1)}$ in parallel. Now x_3 is known and we have $R<n - 2>$ system.
- • Thus, for the k-th step, x_k is known. Compute $c_i^{(k)} = a_{ik}x_k + c_i^{(k-1)}$ in parallel for $k = 1, 2, 3, \ldots, n - 1$, where $c_i^{(0)} = c_i$.

This algorithm needs $n - 1$ processors for $k = 1$ and fewer thereafter. The number of time unit steps for the solution of the $R<n>$ system with $p = n - 1$ processors is $T_p = 2(n - 1)$. $T_1 = n(n - 1)$. Thus the speed-up factor and efficiency are respectively

$$S_p = T_1/T_p = n/2,\; E_p = S_p/p = n/(2(n - 1)) > 0.5.$$

Clearly, the column-sweep algorithm for $R<n, m>$ system has speed-up factor O(m). The $R<N, m>$ system with $m \ll n$ is of interest in practice. Usually m is 1 or 2. However, m is the maximal number of processors used. The column-sweep algorithm is expensive for an SIMD model with a large number of processors.

Linear square system Ax = b We now consider solving linear equations A**x** = **b** using iterative methods, where the matrix A is square. Without any loss of generality, consider the diagonal elements of A to be unity. Write

$$A = A_R + I + A_L,$$

where A_R and A_L are strictly upper and strictly lower triangular matrices and I is the unit matrix of the same order as that of A.

The ***Jacobi iterative algorithm*** for the solution vector **x** is

$$\mathbf{x}^{(k+1)} = -(A_L + A_R)\mathbf{x}^{(k)} + \mathbf{b},\ k = 0, 1, 2, 3, \ldots$$

where $\mathbf{x}^{(0)}$ is a suitably chosen initial approximation. This is a matrix-vector iteration. The algorithm is inherently parallel. The ***Gauss-Seidel iteration algorithm*** is

$$(I + A_R)\mathbf{x}^{(k)} = -A_L\mathbf{x}^{(k-1)} + \mathbf{b},\ \ k = 0, 1, 2, 3, \ldots,$$

where $\mathbf{x}^{(0)}$ is a suitably chosen initial approximation. This algorithm is not parallel in this form. To introduce parallelism, let

$$\mathbf{z} = -A_L\mathbf{x}^{(k-1)} + \mathbf{b},\ \mathbf{y} = \mathbf{x}^{(k)}.$$

Then the elements of **z** can be computed in parallel and the foregoing algorithm can be written as

$$(I + A_R)\mathbf{x}^{(k)} = \mathbf{z}$$

and hence

$$\mathbf{y}^{(i)} = \mathbf{z} - A_R\mathbf{y}^{(i-1)},\ \ i = 1, 2, 3, \ldots,$$

The successive over-/under-relaxation algorithm is

$$(I + A_R)\mathbf{x}^{(k)} = ((1 - w)I - wA_L)\mathbf{x}^{(k+1)} + w\mathbf{b},\ k = 1, 2, 3, \ldots,$$

where $\mathbf{x}^{(0)}$ is a suitably chosen initial approximation. If $0 < w < 1$, then this algorithm is successive underrelaxation. If $w = 1$, then it is Gauss-Seidel. If $1 < w < 2$, then it is successive overrelaxation. The choice of w is important. A proper choice of w will permit this algorithm to converge faster. As before, let

$$\mathbf{z}_w = ((1 - w)I - wA_L)\mathbf{x}^{(k+1)} + w\mathbf{b},\ \mathbf{y}_w = \mathbf{x}^{(k)}.$$

Then the foregoing algorithm reduces to the form

$$(I + wA_R)\mathbf{y} = \mathbf{z}$$

which is solved using the (inner) iteration.

$$\mathbf{y}_w^{(i)} = \mathbf{z}_w - wA_R\mathbf{y}_w^{(i-1)}, \quad i = 1, 2, 3, \ldots.$$

Although the programmer can design and write a parallel program for a given algorithm, there are translators, for example, PTRAN (Parallel Translator), which have been developed and still are being improved to convert sequential unstructured as well as structured programs into parallel programs suited for general parallel architectures (machines). The PTRAN studies control flow as well as data flow of the programs and analyzes control dependence and also data dependence of successive statements/boxes in order to determine the control independence and data independence of the statements; such statements are now marked for parallel execution. PTRAN or similar translators will thus allow already existing unstructured/structured sequential Fortran programs (thousands in number) to be parallelized and used in current parallel machines. This will save enormous human effort in parallelizing algorithms manually and thus the resulting cost. The sequential programs will remain useful. Interestingly, a common human being seems to think sequentially in the conscious state of mind and hence writing a sequential algorithm rather than a parallel algorithm is easier and will not turn out to be obsolete so soon. Furhter, it may be observed that "to err is human" while "not to err is computer". Thus the computer conversion of serial programs into parallel ones will be, in general, extremely quick and mistakeless (assuming the conversion routines are bug-free).

6.4 Error-bounds in probabilistic computation

We consider here two probabilistic algorithms, viz., (i) PAC (Probably Approximately Correct) rectangular[3] concept learning[4] algorithm and (ii) a shrinking-rectangle randomized algorithm with interpolation for a complex zero of a function to discuss the error-bounds in these algorithms. Both these algorithms are polynomial-time as these should be.

---

[3]Rectangular concept is the concept geometrically represented by a rectangle each point of which represents the learnt (or to be learnt) concept. For example, the height-range and the weight range of persons, say 155 cm-175 cm and 55 kg-75 kg could define geometrically the rectangle representing the medium-built persons. A nonrectangular concept, on the other hand, could be that of the border of USA represented by latitudes and longitudes.

[4]Learning is defined as the psychological processes or mechanisms that reflect arelatively permanent or enduring change in immediate or potential behavior that results from specific experience with environmental events. The goal of learning is to behave better as a result of experience. There are four types of learning — supervised, unsupervised, reinforcement, and inductive. The task of pure inductive inference (or induction) in inductive learning is that, given a collection of examples of {x, f(x)}, return a function 'h' that approximates 'f'. The function 'h' is called a hypothesis.

6.4.1 PAC learning model

The way the computer (machine) is made to learn a particular concept, say, the concept of medium-cholesterol persons, is analogous to the one a human being would learn. While a common human being could have the disadvantage of forgetting an already learnt information or of not remembering too many things at a time, the machine does not have such a disadvantage. The probably approximately correct (PAC) learning (Valiant 1984) is not expected to provide with 100% certainty the answer to a concerned question when a novel situation is posed to it after it has learnt the concept reasonably well.

We illustrate the algorithm/model by considering learning the concept of medium-cholesterol[5] persons. We denote by

EX = Instance Space
c = Target Concept (e.g. Medium-cholesterol persons)
  = True Concept we are aiming to learn
  ($c \subseteq$ EX, each subset of EX is a concept)
C = Concept Class (e.g. Cholesterol of Persons) = Class of Concepts
  ($C \subseteq \{c \mid c \subseteq EX\}$
D = Arbitrary Unknown Distribution (with No Restriction) on EX
ε = Accuracy Bound
δ = Confidence Bound
h = Hypothesis of the Target Concept c Based on
  Labelled Examples (Drawn Randomly from D) of EX
  = Concept we have learnt.

4.2 PAC Concept Learning

Suppose the machine wants to learn *medium cholesterol person* from instances. For convenience of the reader, let there be the statistical fact that the free cholesterol (measured in mg/100 ml) in human blood has the observed ranges 12–119 with mean 66 in India[6] . Similarly the ester cholesterol (also measured in mg/100 ml) has the observed range 72–159 with mean 116 in India[7]. These ranges are different for Europe and America. However, the medium cholesterol

---

[5] Cholesterol is a waxy fat-like substance that belongs to a class of molecules called steroids. It is found in animal tissue (body cells) and made in liver and is also created by cells lining the small intestine and by individual cells in the body. An excess of free (i.e., low-density lipoprotein — LDL) cholesterol may cause heart disease and stroke. However, cholesterol is essential for (i) formation (and maintenance) of cell membranes, sex hormones (progesterone, testosterone, estradiol, cortisol), (ii) production of bile salts, that help to digest food, and (iii) conversion into vitamin D when exposed to sunlight.

[6] The corresponding range for Europe and America is 45-61.

[7] The corresponding range for Europe and America is 105-189.

person is defined as one whose free cholesterol lies in [50, 90] and ester cholesterol lies in [100, 140]. ***These foregoing information is completely unknown to the machine/us.*** Machine will choose randomly at a time one labelled instance consisting of (i) free cholesterol level, (ii) ester cholesterol level, and (iii) the information whether the person is medium cholesterol or not. The instance space consisting of m instances would look like the one in the following table (Table 6.3).

Table 6.3. Instance space (Training set) EX

Instance Label	Cholesterol (Free)	Cholesterol (Ester)	Cholesterol Person
1	87	130	Medium
2	92	135	Not Medium
3	71	111	Medium
⋮	⋮	⋮	⋮
m	45	90	Not Medium

We would like to learn the concept "medium-cholesterol person" from instances in the instance space called the training set. We are told for each [free cholesterol, ester cholesterol] pair if it is or not of medium-cholesterol. We would like to learn the rectangular concept, i.e., produce an algorithm that in the future answers correctly if a pair [free cholesterol, ester cholesterol] represents not a medium-cholesterol person. We are interested in knowing which value of m to use if we want to learn the concept ***well***. It is necessary to quantize/characterize what we mean by ***well*** when evaluating learnt concepts. This characterization necessitates specifying the ***probability of error of the learnt concept as at most ε.***

The problem is as follows. Learner is presented with C, ε, and δ. He will now formulate/produce h of c based on randomly drawn labelled examples of EX (Figure 6.4).

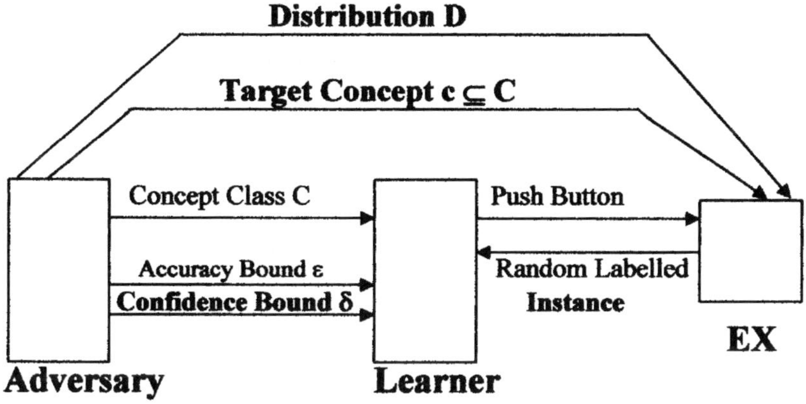

Figure 6.4: PAC learning model

Since c is the true concept we are learning and h is the concept we have learnt, we have

$$\mathrm{Error}_D(h) = P(c \,\Theta\, h) = \sum_{c(x) \neq h(x)} P[x], \quad x = \text{an individual}$$

The goals of the learner are

(i) With high probability (confidence) ($\geq 1 - \delta$) $h \approx c$, i.e., $\mathrm{Error}_D(h) \leq \varepsilon$,

(ii) Time complexity and sample complexity (number m of the labeled examples needed by the algorithm) are polynomial in the size of c, $1/\varepsilon$, $1/\delta$.

A learnt concept h is good when

$P[\mathrm{Error}_D(h) > \varepsilon] < \delta$, i.e., $P[\mathrm{Error}_D(h) \leq \varepsilon] \geq 1 - \delta$.

Degrees of goodness correspond to different values of ε and δ. If

$\varepsilon \to 0$, $\delta \to 0$ then h (learnt concept) $\to$ c

This method of evaluating learning is ***PAC learning***.

6.4.3 Problem on "Learning rectangular concepts" and solution

Given c (= medium-cholesterol persons), ε, δ, to determine the size of the training set we asssume

- c is represented as a rectangle with sides parallel to the axes "free cholesterol/ester cholesterol" with dimensions free-cholesterol-min, free-cholesterol-max, ester-cholesterol-min, ester-cholesterol-max.
- h also will take the form of a rectangle with the sides parallel to the axes.

6.4.4 Algorithm to build h from the training set

- Choose ε, δ.
- An individual x will be classified incorrectly by h if x lies in the area between h and c. Divide this area into 4 strips – on top, bottom, and sides of h (Figure 6.5).

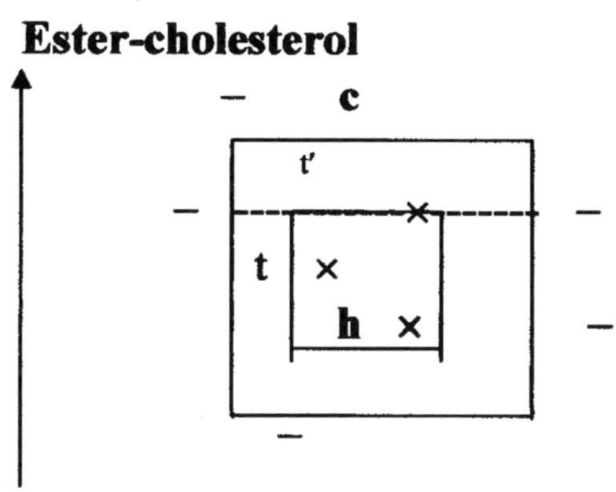

Figure 6.5: Learning medium-cholesterol persons

- Allow these 4 strips to overlap at the 4 corners.
- If area of each strip $\leq \varepsilon/4$ i.e. if this area is contained in the strip t of area $\varepsilon/4$ then

Error (h) $\leq \varepsilon$ (valid for any chosen D).

P(one individual is outside one strip t) $= 1 - \varepsilon/4$.
P(m individuals are outside one strip t) $= (1 - \varepsilon/4)^m$.
P(m individuals are outside at least one of the 4 strips) $= 4(1 - \varepsilon/4)^m$.

Observe that when all the individuals are outside at least one of the strips then the probability of error for an individual > ε (pessimistic). If we bound

$$4(1 - \varepsilon/4)^m < \delta,$$

i.e. $m \ln (1 - \varepsilon/4) < \ln (\delta/4)$,

i.e. $m > \ln (\delta/4)/ \ln (1 - \varepsilon/4)$ since $\ln (1 - \varepsilon/4) < 0$,

i.e. $m > (-4/\varepsilon) \ln (\delta/4)$ or, equivalently, $m > (4/\varepsilon) \ln (4/\delta)$,

we get the lower bound on m. Some representative values of m are given in Table 6.4.

Table 6.4 Number of examples m for specified error ε and (lack of) confidence δ

ε	δ	m
.1	.1	148
.1	.01	240
.1	.001	332
.01	.1	1476
.01	.01	2397
.01	.001	3318
.001	.1	14765
.001	.01	23966
.001	.001	33176

6.4.5 Subcases of learning "Rectangular concepts"

The following subcases are of interest. These are

(i) If two of the sides overlap the axes, we have 2 strips only.
(ii) If the concept is 1-D interval (say, medium-ester-cholesterol person), we still have 2 strips.
(iii) If the concept is 1-D interval starting with 0, we have 1 strip.

6.4.6 Problems

We have now the following problems.

Problem 1 How do you choose the size of the training set so that we have confidence $1 - \delta$ that

$$\text{Error}_D(h) < \varepsilon\ ?$$

Problem 2 Let the concept class C = {set of all polynomials p(x)}. Let the instance space EX consist of values of x and corresponding p(x). If $\varepsilon = .1$ and $\delta = .1$ then what would be the size m of the training set to learn the target concept of a polynomial?

6.4.7 Remarks

The computational learning theory (CLT) is an intersection of artificial Intelligence (AI) and theoretical computer science (TCS), i.e., CLT = AI $\cap$ TCS while PAC learning $\subset$ CLT. In CLT, learning is defined as behaving better as a result of experience. We have the following unanswered questions.

Q. 1 How can we know that one's learning algorithm has produced a theory that will correctly predict the future?

Q. 2 How do we know that the hypothesis h is close to the target concept/function c if we do not know what c is?

However, any hypothesis h that is consistent with a sufficiently large number of training examples (randomly taken from EX) is unlikely to be seriously wrong i.e. it must be PAC.

Further, any h that is seriously wrong will almost certainly be found out with high probability after a small number of examples because it makes an incorrect prediction.

Stationarity Assumption (Valiant 1984) Both the training set and the test set are drawn randomly from the same population of examples i.e. Instance Space EX using the same probability distribution. This assumption makes necessary connection between future and past.

How many examples are needed ? i.e. m = ? The bound on m has already been derived using error in h involving ε and δ. ***If we want ε=0 and δ=0 then m will be infinity.***

If C = {set of all Boolean functions on n attributes} then the size of C =|C| =2^2^n (i.e., 2 to the power of 2 to the power of n). Thus the sample complexity of the hyperspace grows as 2^n. Since the number of +ve examples is also 2^n, any learning algorithm for the space of all Boolean functions will do no better than a look-up table, if it merely returns a hypothesis that is consistent with all known examples.

Dilemma If we do not restrict the space of concept class/functions C the algorithm can consider then it will not be able to learn. If we restrict the space then we may eliminate the true (target) concept/function altogether.

6.5 Shrinking-rectangle randomized algorithm for complex zero: Error and complexity

6.5.1 Introduction

A polynomial-time deterministic randomised algorithm (Sen and Sen 2002) is described to compute a zero of a complex/real polynomial or a complex/real transcendental function in a complex plane. The algorithm starts with a specified rectangle enclosing a complex zero, shrinks it successively by at least 50% in each iteration somewhat like a two-dimensional bisection, and then a single application of linear two-variable interpolation in the highly shrunk rectangle provides the required zero. A parallel implementation of this algorithm is discussed while its sequential and parallel computational complexities as well as its space complexity are presented. The algorithm is found to be reasonably good in terms of the quality of the result for zero clusters and also for multiple zeros. This method can be extended to minimize globally a polynomial or a transcendental function of several variables without resorting to the computation of its partial derivatives and can be used along with the deflation of the polynomial or with different specified initial rectangle.

There exist several deterministic nonrandomized polynomial-time algorithms in the literature (Krishnamurthy and Sen 2001, Mathews 1994, Schilling and Harries 2002) to compute real and complex roots of an algebraic or a transcendental equation. Specifically, automatic two-dimensional (2-D) bisection methods (Sen and Lord 1990, Wilf 1978) have been described. These bisection methods needing only function computations provide excellent accuracy for well-conditioned functions, i.e., functions without zero-clusters (closely spaced zeros). A function with distinct well-separated zeros or one with multiple zeros is well-conditioned with respect to its zero-finding using a bisection method. We present here a variation of the 2-D bisection method, called here the SRA algorithm, that differs widely from these methods in its conceptual approach. We specify a reasonably small rectangle in the complex plane so that it contains one of the complex zeros (including multiple zeros) of the given one-variable function. We throw uniformly distributed darts, say 10 or 20 of them, onto the rectangle and compute at each point of hit the value of the function. This dart-throwing is accomplished by generating uniformly distributed pairs of pseudo-random numbers each representing a point in the rectangle. We select the point of hit, that corresponds to the minimum value of the norm of the function; replace the rectangle by one whose area is at least 50% less than the previous one. This smaller rectangle now has in it the zero of the function. We continue this process a number of times, say 10 times, and obtain the highly shrunk rectangle which contains this zero. Carry out a two-variable linear interpolation to obtain the zero with sufficient accuracy. To obtain other complex zeros of the function, one can deflate the function (algebraic) and follow the proposed algorithm on the deflated function or one can choose a different rectangle that would contain the second zero of the function (algebraic or trigonometric) and apply the algorithm. This algorithm can be modified for a function of several variables by generating a set of several pseudorandom numbers instead of the pair and by using multi-variable interpolation to obtain a global minimum of the function. We will not discuss this aspect here.

The algorithm with the justification of each of its steps is described in Section 6.5.2 while its computational and space complexities are discussed in Section 6.5.3. A Matlab program that not only takes care of complex zeros with nonzero real and nonzero imaginary parts but also purely real as well as purely imaginary zeros is presented in Section 6.5.4. Test examples and conclusions are included in Sections 6.5.5 and 6.5.6, respectively.

6.5.2 The SRA algorithm

Let f(x) be an algebraic or a transcendental function of the single variable x and be continuous and bounded. The function f(x) may be real or complex. The problem is to find a complex zero of f(x), where the zero may have (i) both real and imaginary parts nonzero or may be (ii) only real or (iii) only imaginary. It can be seen that the two-variable interpolation needed for Case (i) is not applicable to any of Cases (ii) and (iii). In the later two cases, we have to explicitly use distinct single variable interpolations.

Let the real part of the zero of the continuous bounded function f(x) lie in the interval [a, b] while the imaginary part in the interval [c, d]. These intervals define a rectangle D (Figure 6.6a), i.e. the domain of search. The function values corresponding to each of the points of the domain will define a plane (i.e., a hyperplane of dimension 2) R whose sides are in general curved (not straight lines) (Figure 6.6b). Assume that this domain contains only one zero for the sake of observing how the algorithm works, such an assumption is not strictly needed though. This fact will be seen when we consider numerical test examples that illustrate the algorithm.

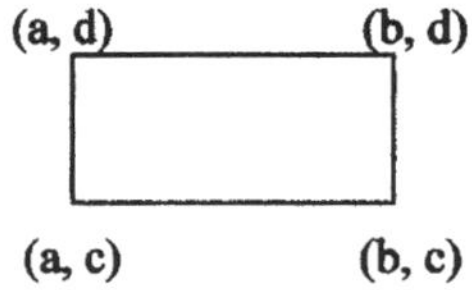

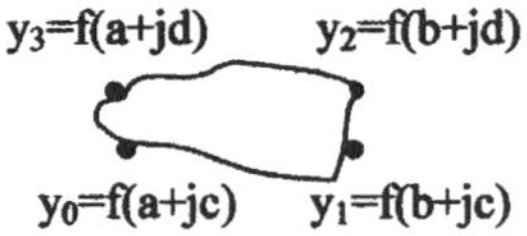

Figure 6.6a: Rectangle D containing a zero of a function (Domain space)
Figure 6.6 b: Plane R with curved sides (Function space)

We now describe the steps of the algorithm with justification/explanation for each step.

S. 1 ***Throwing darts onto the rectangle* D *and choosing the one nearest zero*** Generate uniformly distributed n_1 (n_1 = 20, say) ordered pairs of random numbers such that the first number of each pair lies in [a, b] while the second lies in [c, d]. Each pair defines a complex random number. The first one of the pair defines the real part while the second one the imaginary part of this complex random number. Let rand(1) produce a random number in (0, 1). Then $x = [\text{rand}(1)(b - a) + a] + j\,[\text{rand}(1)(d - c) + c]$ is a complex random number inside the rectangle D. Note that the second Matlab rand(1) produces a random number usually different from that produced by the first rand(1). This number corresponds to a thrown dart inside D and the corresponding f(x) will have a numerical value which is usually complex and which will lie inside the plane R. Corresponding to n_1 such complex random numbers within the rectangle D, we will have n_1 function values within the plane R. We choose that random number which produces the minimum absolute value of the function. Call this number x_1. Clearly this absolute value of $f(x_1)$ will be nearest the zero of f(x).

S. 2 *Shrinking the Rectangle D by at least 50%* Set $b_1 := \text{real}(x_1) + .354(b - a)$, $a_1 := \text{real}(x_1) - 0.354(b - a)$, where $\text{real}(x_1)$ is the real part of x_1. If $b_1 < b$ then set $b := b_1$; if $a_1 > a$ then set $a := a_1$. Similarly, set $d_1 := \text{imag}(x_1) + .354(d - c)$, $c_1 := \text{imag}(x_1) - (d - c)$, where $\text{imag}(x_1)$ is the imaginary part of x_1. If $d_1 < d$ then set $d := d_1$; if $c_1 > c$ then set $c := c_1$.

The step S. 2 reduces the rectangle D by at least half its size. The new rectangle will enclose the zero of f(x) assuming that it is not too violently fluctuating or the zeros are not too closely spaced.

S. 3 *Getting the smallest rectangle after k iterations* Repeat the steps S. 1 and S. 2 for k (k = 10, say) times. This step will produce a highly shrunk rectangle that contains the zero of f(x).

S. 4 *Two-variable interpolation for a complex zero* Use the two-variable Lagrange linear interpolation using the most recent values of a, b, c, d and the corresponding function values. This interpolation includes extrapolation automatically. Let (x_i, y_i) $i = 0(1)3$ be the table for interpolation, where x_i as well as y_i are both complex and the interpolation (that includes extrapolation too) problem is posed as follows.

x	$x_0 = a + jc$	$x_1 = b + jc$	$x_2 = b + jd$	$x_3 = a + jd$	$x = ?$
y	$y_0 = f(x_0)$	$y_1 = f(x_1)$	$y_2 = f(x_2)$	$y_3 = f(x_3)$	$y = f(x) = 0$

Hence, if $a \neq 0$, $b \neq 0$, $a \neq b$, $d_1 = y_0 - y_1 \neq 0$, $d_2 = y_0 - y_2 \neq 0$, $d_3 = y_0 - y_3 \neq 0$, $d_4 = y_1 - y_2 \neq 0$, $d_5 = y_1 - y_3 \neq 0$, $d_6 = y_2 - y_3 \neq 0$, $d_7 = y_1y_2$, $d_8 = y_1y_3$, $d_9 = y_2y_3$, then

$$x = -\,x_0y_1d_9/(d_1d_2d_3) + x_1y_0d_9/(d_1d_4d_5) - x_2y_0d_8/(d_2d_4d_6) + x_3y_0d_7/(d_3d_5d_6) \quad \ldots \quad (6.1)$$

This interpolation is carried out only once in the final highly shrunk rectangle. The x thus obtained is the required zero of the function f(x).

Interpolation for computing only a real zero The foregoing interpolation formula (1) is not valid for obtaining a real zero of f(x) since $y_0 = y_3$ and $y_1 = y_2$ and consequently d_3 and d_4 both are zero and each one occurs in the denominator in the formula (1). Therefore, we use the modified interpolation formula

$$x = -x_0y_1/d_1 + x_1y_0/d_1 \quad \textit{(for real zeros only)} \tag{6.2}$$

Interpolation for computing only an imaginary zero The formula (6.1) is invalid here too. The modified interpolation formula is

$$x = -\, x_0y_3/d_3 + x_3y_0/d_3 \; \textit{(for imaginary zeros only)} \tag{6.3}$$

The x that we obtain in the formula (6.1) or (6.2) or (6.3) is the required solution. The corresponding function value f(x) will be sufficiently small so that the zero x could be accepted as the required zero for all practical purposes.

S. 5 ***Error in (quality of) the zero x*** "How good is the quality of the zero?" is a very pertinent question that is almost always asked. The answer is obtained through computing a relative error (i.e., error-bound) in the zero x. Observe that an absolute error is not much meaningful in numerical computation. In the absence of the knowledge of the exact zero (solution) which is never known (for if it is numerically known then we do not bring error unnecessarily into the scene), we consider usually the solution (zero) of higher order accuracy for the exact solution. Thus the error in the solution of lower order accuracy will be computed, denoting the solution of higher order accuracy = x_h and the solution of lower order accuracy = x_t, as

$$E_r = (x_h - x_t)/x_h \tag{6.4}$$

Clearly $|f(x_h)| < |f(x_t)|$ by at least an order (Sen 2002). If we consider the interpolated zero (solution) x as the zero (x_t) of lower order accuracy then we do not have the zero (x_h) of higher order accuracy. To determine x_h, we shrink the already highly shrunk rectangle ***once more*** and carry out the interpolation as in the step S. 4. This interpolated zero will be the zero (x_h) of higher order accuracy. Thus we can compute the relative error E_r. The step S. 5 has not been included in the MATLAB program for physical conciseness and for better comprehension. The reader may achieve this step of error computation by running the program for the second time replacing k by k + 1 and obtaining the zero x_h of higher order accuracy. Otherwise, he may automate the program by appropriately modifying it.

6.5.3 Computational and space complexities

The ***computational complexity*** of the SRA algorithm can be derived as follows. To generate n_1 pairs of random numbers using the multiplicative congruential generator or, equivalently, the power residue method (Banks et al. 1998), we need $2n_1$ multiplications and $2n_1$ divisions (to carry out mod operations). To obtain n_1 complex random numbers in the specified rectangle D (Figure6.1a), we need further $2n_1$ multiplications and $2n_1$ additions. If we do not distinguish between a division and a multiplication then so far we need $6n_1$ ***real multiplications*** and $2n_1$ ***real additions*** for generating n_1 complex random numbers. If the function f(x) is a ***polynomial of degree*** n, then the computation of f(x) using the nested multiplication scheme (Krishnamurthy and Sen 2001) would need n complex multiplications and n complex additions, i.e., 2n real multiplications and 2n real additions for each complex random number. Hence, for n_1 complex random numbers, we need $2n \times n_1$ ***real multiplications*** + $2n \times n_1$ ***real additions***. Since we have k rectangles before we reach the smallest one we need, for the computation of the ***smallest rectangle***, $6k \times n_1 + 2k \times n \times n_1$ ***multiplications*** and $2k \times n_1 + 2k \times n \times n_1$ ***additions***. Since k, n_1 are independent of the size n of the function f(x), our computational complexity will $\mathbf{O(2k \times n_1 \times n)}$ assuming n very large (compared to n_1 and k, and the size of the program) but finite. A typical value of k is 10 and that of n_1 is 20. These values, however, will be larger if the initial rectangle chosen is larger.

The ***space complexity***, i.e., the storage space needed to store the input data, viz., the (n + 1) complex coefficients of the nth degree polynomial f(x), we need 2n locations. We also need the storage space to store the program. Since the storage space for the program is independent of the size, i.e., the degree n of f(x), the space complexity is simply **O(2n)** assuming n very large but finite.

If the function f(x) is a ***transcendental function*** then the computational complexity will be O(2k × n_1 × number of operations needed to compute f(x)) while the space complexity will be the space needed for the function. Observe that the transcendental function though may be written as a polynomial of degree $\propto$, does not have the computational complexity O($\propto$) nor has the space complexity O($\propto$).

These complexities are comparable with those of other existing methods. The space complexity as well as the computational complexity in terms of the input size n for all these methods will not be usually $O(n^s)$, where $s > 1$.

The ***parallel computational complexity*** using n processors will clearly depend only on the values of n_1 and k. If we use $p < n$ processors then the complexity will increase proportionately. The space complexity, however, will remain unchanged.

6.5.4 MATLAB program for the SRA algorithm

This program is self-explanatory and computes a complex zero of a polynomial or a transcendental function.

```
function[]=func2(rmin, rmax, imin, imax, nmax, eps, fun)
%func2 computes a complex zero of a function fun
%using a randomized algorithm with an interpolation

%Description of input parameters rmin, rmax, imin, imax, etc.

%[rmin, rmax]=interval of real part of the zero.
%[imin, imax]=interval of imaginary part of the zero.
%nmax=maximum no, of bisections (nmax=10 usually;
%for better accuracy, nmax may be taken as 20 0r 30.
%eps=.5*10^-4 usually; for better accuracy, eps=.5*10^-8.
%However, eps is used here as a relative error term and
%should be chosen compared to the input quantities involved.
%fun is the function, one of whose zeros is to be obtained.
%For example, fun='x^2+x+1' for the function f(x)=x^2+x+1.

for k=1:10
  %This number 10 implies that the original rectangle is
  % shrunk successively 10 times. This number seems reasonably
  %good; however, it may be increased depending on the accuracy
  % needed within the limit of the precision of the computer.

xvect=[];fvect=[]; absfvect=[];
for i=1:nmax
  x=(rand(1)*(rmax-rmin)+rmin)+j*(rand(1)*(imax-imin)+imin);
  f=eval(fun); absf=abs(f);
  xvect=[xvect;x];
  fvect=[fvect;f];
  absfvect=[absfvect; absf];
end;

x_f_absf=[xvect fvect absfvect];
x_f_absf_s=sortrows(x_f_absf, 3);
string 'sorted x, f(x), absolute f(x)'
x_f_absf_s
if abs(x_f_absf_s(1,3))<eps
  string 'root, function-value, absolute function value'
  x_f_absf_s(1,:)
  break
```

```
end;

x1=x_f_absf_s(1,1);
realdiff=rmax-rmin; imagdiff=imax-imin;
rmax1=real(x1)+0.354*realdiff; rmin1=real(x1)-0.354*realdiff;
if rmax1<rmax
  rmax=rmax1;
end;
if rmin1>rmin
  rmin=rmin1;
end;

  imax1=imag(x1)+0.354*imagdiff; imin1=imag(x1)-0.354*imagdiff;
  if imax1<imax
    imax=imax1;
  end;
  if imin1>imin
    imin=imin1;
  end;
  string 'rmax,rmin,imax,imin'
  rmax,rmin,imax,imin
end;

a=rmin; b=rmax; c=imin; d=imax;

%The foregoing statements reduce the rectangle to maximum half its size.
%This reduction has resemblance with 2-D bisection for a complex zero.

  x=a+j*c; x0=x; y0=eval(fun); x=b+j*c; x1=x; y1=eval(fun);
  x=b+j*d; x2=x; y2=eval(fun); x=a+j*d; x3=x; y3=eval(fun);
  d1=y0-y1; d2=y0-y2; d3=y0-y3;d4=y1-y2; d5=y1-y3;d6=y2-y3;

  d7=y1*y2; d8=y1*y3; d9=y2*y3;

  if abs(d1)<eps, d1=1; end; if abs(d2)<eps, d2=1; end;if abs(d3)<eps, d3=1;
end;
  if abs(d4)<eps, d4=1; end; if abs(d5)<eps, d5=1; end;if abs(d6)<eps, d6=1;
end;

    xx0=-x0*y1*d9/(d1*d2*d3);
    xx1=-x1*y0*d9/(-d1*d4*d5);
    xx2=-x2*y0*d8/(d2*d4*d6);
    xx3=-x3*y0*d7/(-d3*d5*d6);
```

```
    if abs(c)<eps & abs(d)<eps, xx0=-x0*y1/d1; xx1=x1*y0/d1;xx2=0;
xx3=0; end;
    %This statement is for interpolation for only real zeros.

    string 'x0, yo, x3,y3,d3'
    % Imaginary x0 & x3 and corresponding y0 & y3 for linear interpola-
tion
    x0,y0,x3,y3,d3

    if abs(a)<eps & abs(b)<eps, xx0=-x0*y3/d3;xx3=x3*y0/d3;xx1=0;
xx2=0;  end;
   %This statement is for inperpolation for only imaginary zeros.

    x=xx0+xx1+xx2+xx3;
  f=eval(fun); absf=abs(f);

  string 'interpolated (including extrapolated) zero, f-value, abs f-value'
  x, f, absf

 if absf<eps
  string 'root,f-value, abs_f-value (correct up to 1/eps digits)'
  x, f, absf
  break
end;
```

6.5.5 Test examples

To check the SRA algorithm, we have constructed several typical test functions (i.e., functions whose zeros are known through the MATLAB function **poly**). To conserve space we present here just four examples.

Example 1 (*A real quatric polynomial with only real zeros*) $f(x) = x^4 - 5.2x^3 + 10.04x^2 - 8.528x + 2.688$ whose exact zeros are 1, 1.2, 1.4, and 1.6 and which is constructed using the MATLAB command **poly([1 1.2 1.4 1.6])**. The inputs are

```
rmin=0;rmax=1.19;imin=0;imax=0;nmax=10;eps=.5*10^-8;fun='x^4-
5.2*x^3+10.04*x^2-8.528*x+2.688';
func2(rmin,rmax,imin,imax,nmax,eps,fun)
```

The outputs are **x = 1.1998, f = -3.0969e-006, absf = 3.0969e-006**. The second run with the same inputs resulted in the outputs **x = 1.0016, f = -7.5359e-005, absf = 7.5359e-005**

Example 2 (A quatric real polynomial having only imaginary zeros) $f(x) = x^4 + 5x^2 + 4$ whose exact zeros are – i, i, – 2i, and 2i. The inputs are

rmin=0;rmax=0;imin=-1.5;imax=-.5;nmax=10;eps=.5*10^-4;fun='x^4+5*x^2+4';
func2(rmin,rmax,imin,imax,nmax,eps,fun)

The outputs are **x = 0 - 1.0000i, f = -1.4188e-004, absf = 1.4188e-004.**

Example 3 (A quatric complex polynomial with zero-clusters: a highly ill-conditioned problem) $f(x) = x^4 - (8.04 + .22j)x^3 + (24.2227 + 1.3266j)x^2 - (32.410446 + 2.665828j)x + (16.25009862 + 1.78524984j)$ whose exact zeros are 2.01 + j.04, 2.01 +j.05, 2.01 + j.06, and 2.01 +j.07, where $j = \sqrt{-1}$. The inputs are

rmin=2;rmax=2.019;imin=0;imax=0.045;nmax=10;eps=.5*10^-8;
» fun='x^4-(8.04+.22*j)*x^3+(24.2227+1.3266*j)*x^2-(32.410446+2.665828*j)*x+(16.25009862+1.78524984*j)';
func2(rmin,rmax,imin,imax,nmax,eps,fun)

The outputs are **x = 2.0112 + 0.0470i, f = -6.5411e-009 -2.3059e-009i,** absf = 6.9356e-009. When the program was rerun with the same inputs, the outputs became x = 2.0110 + 0.0519i, f = 3.5115e-009 -1.4762e-009i, absf = 3.8092e-009. The foregoing results seem reasonably good for the precision of 15 digits that MATLAB provides.

Examples 4 (A tenth degree real polynomial with large coefficients and distinct real zeros) $f(x) = x^{10} - 55x^9 + 1320x^8 - 18150x^7 + 157773x^6 - 902055x^5 + 3416930x^4 - 8409500x^3 + 12753576x^2 - 10628640x + 3628800$ whose zeros are 1, 2, 3, 4, 5, 6, 7, 8, 9, and 10. *The* inputs are

» func2(.1,1.5,0,0,10, .5*10^-8,'x^10-55*x^9+1320*x^8-18150*x^7+157773*x^6-902055*x^5+3416930*x^4-8409500*x^3+12753576*x^2-10628640*x+3628800')

The outputs are **x = 1.0000, f = -9.4986, absf = 9.4986**. When the program was rerun with the same polynomial with changed inputs

» func2(.1,11,0,0,10, .5*10^-8,'x^10-55*x^9+1320*x^8-18150*x^7+157773*x^6-902055*x^5+3416930*x^4-8409500*x^3+12753576*x^2-10628640*x+3628800')

we obtained the outputs **x = 6.0000, f = -0.0570, absf = 0.0570**. When we reran the program with the foregoing inputs for the second, third, fourth, and fifth times then the outputs became (a) **x = 5.9977, f = -6.5615, absf = 6.5615**, (b) **x = 4.0030, f = 12.9012, absf = 12.9012**, (c) **x = 5.9988, f = -3.5642, absf = 3.5642**, (d) **x = 4.0015, f = 6.6179, absf = 6.6179**, respectively.

6.5.6 Remarks

Shrinking rectangle converges faster than 2-D bisection. When k goes to 10, the initial rectangle D (Figure 6a) that encloses/contains a zero of the function f(x) will be shrunk to the rectangle whose area will be less than or equal to $D/2^k = D/2^{10} = 0.00097656D$. This shrinking is significantly rapid compared to the automatic bisection for complex zeros (Sen and Lord 1990, Wilf 1978).

The non-existence of a zero in the wrongly chosen initial rectangle can be detected. The SRA algorithm will come out indicating that the chosen rectangle D does not contain a zero if the choice is incorrect, i.e., if it really does not contain a zero.

Interpolation (including extrapolation) is carried out in the final highly shrunk rectangle only once It is possible to interpolate linearly in each of the k (= 10) rectangles. However, it is not done because the linear interpolation could be sufficiently inaccurate when the rectangle is large. Moreover, such repeated interpolations will not only increase the computation but also might result in excluding the actual zero in the rectangle-shrinking process.

The zero existing in the initial rectangle D will exist in the final shrunk rectangle. In our numerical experiment with numerous functions and with reasonably chosen initial rectangle D, the zero that was located in D always remained in the final shrunk rectangle. The SRA algorithm thus seems an efficient fail-proof complex zero finding method and it is ***deterministic***.

The SRA algorithm is not worse than most algorithms for finding a zero in a zero-cluster. A function having zero-clusters (closely spaced zeros) is always an ill-conditioned problem with respect to finding a zero accurately in the cluster. Any method so far existing as well as any method that could be proposed in future would be only satisfactory to a varying extent for a specified precision. Our numerical experiment depicts that the SRA algorithm is reasonably good when dealing with zero-clusters.

Multiple zeros do not pose any problem to the SRA algorithm. Unlike the Newton method and its variations which need to compute derivatives of a function and in which an oscillation around a multiple zero (in a finite precision machine) sets in, the SRA algorithm has absolutely no such problem. It gives, like bisection methods, the multiple zero accurately as it does not depend on the computation of the derivatives of a function. For a polynomial having multiple zeros, ***repeated deflations*** will provide the order of multiplicity.

Use deflation or different rectangles to seeve out all the zeros. One way of seeving **all the zeros** of a polynomial with or without multiple zeros is to deflate the polynomial successively after computing a zero. The other way is to choose different appropriate intervals/rectangles each enclosing a zero and compute all the zeros. For a transcendental function that cannot be written as the product of a polynomial (with multiple zeros) and another transcendental function, deflations may not be useful.

The SRA algorithm has a sequential complexity O(n) and its parallel implementation is straight-forward. As we have seen in Section 6.5.3 that the SRA algorithm has a sequential computational complexity $O(2k \times n1 \times n)$ where the input size is O(2n) for an nth degree complex polynomial. Observe that k (=10, say) and n_1 (=10 or 20, say) are independent of n. The parallel computational complexity, when we have n processors, is $O(k \times n_1)$ which is independent of the input size. For a fixed number of processors < n, this complexity will increase proportionately.

The SRA algorithm can be extended to obtain the global minimum of a multi-variable function. Instead of generating a pair of pseudorandom numbers for a complex zero of a function f(x), we have to generate an ordered set of pseudorandom numbers for this purpose and suitably modify this algorithm.

Bibliography

Baker, G.L.; Gollub, J.P. (1996): *Chaotic Dynamics: an Introduction*, Cambridge University Press.

Banks, J.;Carson, J.S., II; Nelson, B.L. (1998): *Discrete-event Simulation* (2nd ed.), Prentice-Hall of India, New Delhi.

Cellier, F.E. (1998): *Continuous System Simulation*, Springer-Verlag, New York.

Gregory, R.T.; Krishnamurthy, E.V. (1984): *Methods and Applications of Error-free Computation*, Springer, New York.

Jain, M.K.; Ramful, A.; Sen, S.K. (2000): Solving linear differential equations as a minimum norm least squares problem with error-bounds, *Intern. J. Computer Math.*, **74**, 325-343.

Karmarkar, N. (1984): A new polynomial-time algorithm in linear programming, *Combinatorics*,**4**, 373-395.

Khachian, L.G. (1979): A polynomial algorithm in linear programming, *Dokl. Akad. Nauk USSR*, 244, 1093-1096, translated as *Soviet Math. Dokl.* 20, 191-194.

Krishnamurthy, E.V.; Sen, S.K. (2001): *Numerical Algorithms: Computations in Science and Engineering*, Affiliated East-West Press, New Delhi.

Kulisch, U.W.; Miranker, W.L. (1986): The arithmetic of the digital computer: a new approach, *SIAM Review*, 28, 1-40.

Lakshmikantham, V.; Sen, S.K.; Sivasundaram, S. (1995): Computing polynomial root-clusters exactly and parallely, *Engineering Simulation*, 12, 1995, 291 – 313.

Lakshmikantham, V.; Sen, S.K.; Howell, G. (1996): Vectors versus matrices: p-inversion, cryptographic application, and vector implementation, *Neural, Parallel & Scientific Computations,* **4**, 129-140.

Lakshmikantham, V.; Maulloo, A.K.; Sen, S.K.; Sivasundaram, S. (1997): Solving linear programming problems exactly, *Applied Mathematics and Computation*, 1997, **81**, 69-87.

Lakshmikantham, V.; Sen, S.K.; Jain, M.K.; Ramful, A. (2000): $O(n^3)$ noniterative heuristic algorithm for linear programs with error-free implementation, *Applied Mathematics and Computation*, **110**, 53-81.

Lakshmikantham, V.; Sen, S.K.; Mohanty, A. (2004): Error in error-free computation for linear system, to appear.

Lord, E.A.; Sen, S.K.; Venkaiah, V.Ch. (1990): A concise algorithm to solve under-/over-determined linear systems, *Simulation*, **54**, 239-240.

Lord, E.A.; Venkaiah, V.Ch.; Sen, S.K. (1996): A shrinking polytope method for linear programming, *Neural, Parallel & Scientific Computations,* **4**, 325-340.

Mathews, J.H. (1994): *Numerical Methods for Mathematics, Science, and Engineering,* 2nd ed., Prentice-Hall of India, New Delhi.

Quinn, M.J. (1987): *Designing Efficient Algorithms for Parallel Computers*, McGraw-Hill, Singapore.

Rokne, J.; Lancaster, P., Complex interval arithmetic, *Comm. ACM*, 1971, **14**, 111-112.

Schilling, R.J.; Harries, S.L. (2002): *Applied Numerical Methods for Engineers, using MATLAB and C,* Thomson Asia Pvt. Ltd., Singapore.

Sen, S.K. (2003): Error and Computational Complexity in Engineering, in *Computational Mathematics, Modelling and Algorithms* (Chap. 5) ed. J.C. Misra, Narosa Publishing House, New Delhi, 110-145.

Sen, S.K.; Lord, E.A. (1990): An automatic bisection to compute complex zeros of function, in S. Bandyopadhyay (ed.), *Information Technology: Key to Progress*, Tata-McGraw-Hill, New Delhi, 9-13.

Sen, S.K.; Howell, G. (1992): Direct fail-proof triangularization algorithms for $AX + XB = C$ with error-free and parallel implementations, *J. Appl. Maths. And Computation (Elsevier Science Pub. Co., New York),* 50, 255 – 278.

Sen, S.K.; Maulloo, A.K. (1994): Inequality sorting algorithm with p-adic arithmetic to solve LP problems exactly, *39th Congress of ISTAM*, Andhra University, Waltair, Dec. 27 – 30, 1994, 57 – 58 (abstract).

Sen, S.K.; Mohanty, A. (2003): Error estimate for error-free computation for linear system, *Proc. 48th Congress of ISTAM (International Meet)*, Ranchi, Dec 18-21, 2003, 53-62.

Sen, S.K.; Sen, S.(2002): A shrinking-rectangle randomized algorithm with interpolation for a complex zero of a function, **Proc.** *47th Congress of ISTAM (An International Meet),* Indian Institute of Technology, Guwahati, Dec 23-26, 2002, 72-80.

Sen, S.K.; Sen, S. (2004): $O(n^3)$ g-inversion-free noniterative near-consistent linear system solver for minimum-norm least-squares and nonnegative solutions, to appear in *J. Computational Methods in Sciences and Engineering.*

Traub, J.F.; Wozniakowski, H. (1982): Complexity of linear programming, *Operations Research Letters*, 1, No. 1, 59-62.

Turing, A.M. (1936): On computable numbers with an application to the Entscheidungs problem, *Proc. London Math. Soc.*, 42, (Series 2), 230-65.

Valiant, L.G. (1984): A theory of the learnable, *Comm. ACM,* **27**, No. 11, 1134-42.

Wilf, H. (1978): A global bisection method for computing the zeros of a polynomial in the complex plane*, J. ACM*, 415-420.

Index

Mathematics in Science and Engineering
Edited by C.K. Chui, Stanford University

Recent titles:
I. Podlubny, *Fractional Differential Equations*
E. Castillo, A. Iglesias, R. Ruíz-Cobo, *Functional Equations in Applied Sciences*
V. Hutson, J.S. Pym, M.J. Cloud, *Applications of Functional Analysis and Operator Theory (Second Edition)*